Ocean in the Earth System

From the ***Seas and Oceans*** Set
coordinated by
André Mariotti and Jean-Charles Pomerol

Ocean in the Earth System

Edited by

André Monaco
Patrick Prouzet

iSTE

WILEY

First published 2014 in Great Britain and the United States by ISTE Ltd and John Wiley & Sons, Inc.

ISTE Ltd
27-37 St George's Road
London SW19 4EU
UK

www.iste.co.uk

John Wiley & Sons, Inc.
111 River Street
Hoboken, NJ 07030
USA

www.wiley.com

Library of Congress Control Number: 2014950495

British Library Cataloguing-in-Publication Data
A CIP record for this book is available from the British Library
ISBN 978-1-84821-701-0

Contents

Foreword

We have been asked by ISTE to stimulate work in the area of the environment. Therefore, we are proud to present the "Seas and Oceans" set of books, edited by André Monaco and Patrick Prouzet.

Both the content and the organization of this collection have largely been inspired by the reflection, initiatives and prospective works of a wide variety of national, European and international organizations in the field of the environment.

The "oceanographic" community, in France and internationally – which is recognized for the academic quality of the work it produces, and is determined that its research should be founded on a solid effort in the area of training and knowledge dissemination – was quick to respond to our call, and now offers this set of books, compiled under the skilled supervision of the two editing authors.

Within this community, there is a consensus about the need to promote an interdisciplinary "science of systems" – specifically in reference to the Earth's own "system" – in an all-encompassing approach, with the aim of providing answers about the planet's state, the way it works and the threats it faces, before going on to construct scenarios and lay down the elementary foundations needed for long-term, sustainable environment management, and for societies to adapt as required. This approach facilitates the shift of attention from this fundamental science of systems (based on the analysis of the processes at play, and the way in which they interact at all levels and between all the constituent parts making up the global system) to a "public" type of science, which is finalizable and participative, open to decision-makers, managers and all those who are interested in the future of our planet.

In this community, terms such as "vulnerability", "adaptation" and "sustainability" are commonly employed. We speak of various concepts, approaches or technologies, such as the value of ecosystems, heritage, "green" technologies, "blue" chemistry and renewable energies. Another foray into the field of civilian science lies in the adaptation of research to scales which are compatible with the societal, economic and legal issues, from global to regional to local.

All these aspects contribute to an in-depth understanding of the concept of an ecosystemic approach, the aim of which is the sustainable usage of natural resources, without affecting the quality, the structure or the function of the ecosystems involved. This concept is akin to the "socio-ecosystem approach" as defined by the Millennium Assessment (http://millenniumassessment.org).

In this context, where the complexity of natural systems is compounded with the complexity of societies, it has been difficult (if only because of how specialized the experts are in fairly reduced fields) to take into account the whole of the terrestrial system. Hence, in this editorial domain, the works in the "Seas and Oceans" set are limited to fluid envelopes and their interfaces. In that context, "sea" must be understood in the generic sense, as a general definition of bodies of salt water, as an environment. This includes epicontinental seas, semi-enclosed seas, enclosed seas, or coastal lakes, all of which are home to significant biodiversity and are highly susceptible to environmental impacts. "Ocean", on the other hand, denotes the environmental system, which has a crucial impact on the physical and biological operation of the terrestrial system – particularly in terms of climate regulation, but also in terms of the enormous reservoir of resources they constitute, covering 71% of the planet's surface, with a volume of 1,370 million km^3 of water.

This set of books covers all of these areas, examined from various aspects by specialists in the field: biological, physical or chemical function, biodiversity, vulnerability to climatic impacts, various uses, etc. The systemic approach and the emphasis placed on the available resources will guide readers to aspects of value-creation, governance and public policy. The long-term observation techniques used, new techniques and modeling are also taken into account; they are indispensable tools for the understanding of the dynamics and the integral functioning of the systems.

Finally, treatises will be included which are devoted to methodological or technical aspects.

The project thus conceived has been well received by numerous scientists renowned for their expertise. They belong to a wide variety of French national and international organizations, focusing on the environment.

These experts deserve our heartfelt thanks for committing to this effort in terms of putting their knowledge across and making it accessible, thus providing current students with the fundaments of knowledge which will help open the door to the broad range of careers that the area of the environment holds. These books are also addressed to a wider audience, including local or national governors, players in the decision-making authorities, or indeed "ordinary" citizens looking to be informed by the most authoritative sources.

Our warmest thanks go to André Monaco and Patrick Prouzet for their devotion and perseverance in service of the success of this enterprise.

Finally, we must thank the CNRS and Ifremer for the interest they have shown in this collection and for their financial aid, and we are very grateful to the numerous universities and other organizations which, through their researchers and engineers, have made the results of their reflections and activities available to this instructional corpus.

André MARIOTTI
Professor Emeritus at University Pierre and Marie Curie
Honorary Member of the Institut Universitaire de France
France

Jean-Charles POMEROL
Professor Emeritus at University Pierre and Marie Curie
France

1

The Ocean in the Earth System: Evolution and Regulation

1.1. The Earth system and its components

It is very common today for our planet to be called an "Earth system". The scientific meaning is both a necessary tool of communication and an area of potential confusion. Therefore, let us first endeavor to probe this concept and define how the Earth is a "system" of which the ocean is one of the "components".

A system is classically defined as a group of elements (or components), each of them interacting with the others through certain principles or rules. This definition seems trivial and merits some comments and clarifications.

1.1.1. *A system is a set of objects whose limit is arbitrary, but pertinent*

First, the word "set" implies a grouping within a boundary, defined in a subjective and arbitrary manner by the observer. In what circumstances is the arbitrary choice of a grouping relevant? For the scientific observer, the grouping is most often relevant because it

---

Chapter written by Philippe BERTRAND.

corresponds to the category of "an area of study", generally a characteristic of an academic field. This might be a cell for a cytologist, a multicellular organism for a physiologist, a population or settlement for a population biologist, the ecosystem for an ecologist, the society for a sociologist or for an economist, the area beneath the Earth's surface for a tectonician or volcanologist, the Earth's surface and atmosphere for a climatologist or an oceanographer, etc. For the non-scientific observer, the intuitive grouping is that which corresponds best to a visual entity, whether that be observed by eyesight alone or through images furnished by means of modern technology (from the microscope that allows cells to be seen to satellites that allow the Earth to be seen from space). It is not surprising that the disciplines in the natural sciences, for the most part, are derived from the visual perceptions of the "man on the street" and from technological progress.

The arbitrary identity of a system, through a grouping, is therefore a form of categorization of the type that biologists use for very broad families of life forms (e.g. bacteria and archaea, protist eukaryotes, multicellular or metazoan eukaryotes). We note, however, that this is different from phylogenetic categorization (modern day taxonomy), as this is founded upon similarities in the attributes and in the genetic proximity amongst a range of objects: that of individual living beings.

In the case of a living being, the definition is, in general, relevant because of its physical limits (e.g. cuticles and skin) and its autonomy. Note that the idea of autonomy can after all be very weak. Is an ant autonomous without its anthill? Is a cell autonomous in a multicellular organism? Is a man or woman autonomous on a reproductive level?

The definition of a system is, in retrospect, very arbitrary when the proportion of non-living components of the system is important (e.g. ecosystems and societies). Such a system should therefore be the subject of a precise and rigorous description within the framework of the limits that have been arbitrarily fixed upon it. Too often, the notion of an ecosystem is, unfortunately, employed as a generalization, each user implicitly conferring a different typology on it.

In the extreme case of a planetary system, the limit again becomes pertinent (the summit of the atmosphere, if there is one, or the soil surface, or the sea surface), since there is a great contrast between a medium where, under the effect of gravitational attraction, the density of the matter is high (the interior of the system, atmosphere included) and a medium where the density of matter is close to zero (the planet's exterior, interplanetary space). However, unlike living beings, such a system is not "reproducible" in a biological sense; in other words, it has no equivalents. Despite that, it can, like living beings, have global characteristics essentially invisible to the non-scientific observer, notably in terms of self-organization, regulation and adaptive evolution.

1.1.2. *One system is necessarily built into another*

A multicellular organism is, as the name suggests, an organized living being; in other words, it is composed of organs that interact with each other in that they maintain the existence of the organism as a whole. But the organisms themselves are composed of specialized (somatic) cells that have interactions between them. It is therefore evident that the arbitrary limit of the internal components of a system, which amount to subsystems but not necessarily to organs, depends directly on its external definition, that is to say on the identity that we attribute to it.

The natural systems are therefore like Russian dolls on the small just as on the large scale, the absolute theoretical limits only being imposed on a small scale by: the indivisibility and the indetermination of the most fundamental quantum particles (e.g. quarks), the indivisibility of the information (quantum information, Boltzmann constant h) or the indivisibility of time ("Planck" time = 10^{-43} s). On a large scale, they are imposed by the spatio-temporal frontiers of the observable universe, in the case of the Big Bang, 13.7 billion years ago, and the finite speed of light, places or events that have occurred at a great distance from us. However, there is a theoretical intermediate limit between the ensemble of "macroscopic" components, dominated by interactions that can be described in a mechanistic way (contacts, flux of matter and energy), and that of "microscopic" components, dominated by interactions that can only be described statistically (movements, interactions at a distance and collisions of quantum particles, atoms and molecules). We can, for example, place this limit at the level of biological cells, the cells being in some way the smallest macroscopic systems, but their contents are relevant for much of the microscopic world (strictly speaking, intracellular organelles are also macroscopic and the base sequences at the heart of a gene on a strand of deoxyribonucleic acid (DNA) can be described in a mechanistic way).

Another theoretical intermediate limit distinguishes isolated or closed systems from open systems. It is this point that we will now tackle.

1.1.3. *The Earth is a "closed" system*

In thermodynamics, a system is called "isolated" when it does not exchange energy nor matter with the outside world. The term "matter" is theoretically useless here, when we consider that from a physicist's point of view, according to Einstein's theory of general relativity, matter is only a particular form of energy. This only accounts for the principle of conservation of energy, matter being only an approximation for the scales of time and space relevant to the life of mankind or the history of humanity.

By virtue of the second law of thermodynamics, an isolated system naturally evolves toward a situation of equilibrium, which is known as "thermodynamic death". In our own perceptions, such a system does not exist because any attempt to observe it would, in fact, break its isolation. It is therefore only theoretical and can only give rise to non-refutable hypotheses whose scientific character can be discussed [POP 62] (such as the cosmological hypothesis of multiverses where universes evolve in isolation from one another).

Conversely, a system is called "open" when it exchanges energy and matter with the outside world. In this case, the system can depart from a situation of thermodynamic equilibrium to reach, eventually, a stationary situation without equilibrium (regulation) whose maintenance is assured by the transformation of a flux of energy during its passage to the heart of the system. This transformation is the price to be paid in order for the level of organization of the system to be maintained or increased. This is what occurs during photosynthesis in plants by directly transforming light energy from the Sun, but also more indirectly, during respiration in animals, plants or bacteria. By doing so, these reuse, more or less directly, the chemical energy that the products of the initial photosynthesis contain; that is to say organic matter (reduced carbon) and oxygen (O_2).

A stable system does not accumulate energy in the long-term. Apart from a few insignificant and temporary fluctuations, such as phases of lipid stocking and the development of living creatures, the entirety of the adsorbed energy is sent back outside the system, only with a modification in quality (an increase in entropy). This is, for example, the case with changes in quality of the energy contained in organic matter (chemical energy) which, after being used by a living consumer (via respiration, by using oxygen or another oxidizing compound), is partly transformed into heat that is then transferred into the surrounding fluid. In this sense, this flux of energy can be considered a mediator that reduces the entropy to the system. As entropy is a physical concept of the measurement of chaos, this reduction results in the system maintaining or increasing its “order”, otherwise known as its level of organization.

The Earth, during its accretion phase (around 4.5 billion years ago), which lasted around 100 million years, was a largely open system which received a very significant influx of matter relative to the mass already accumulated. It was in the next phase, during this continued influx of matter, that terrestrial water, which we know today in the form of oceans, ice, rivers, lakes, groundwater, clouds and vapor, accumulated on the new planet. Simple organic molecules, formed in space, also arrived on Earth during this period. It remained a very open system around 4 billion years ago during a phase of late and intense meteoric bombardment.

But since this very distant epoch, the Earth has not experienced any further significant accumulation of matter. Of course, the current average inward and outward fluxes of matter, at the boundary of the Earth system, are not null – in the order of several hundreds of thousands of tons per year – but they are entirely negligible compared to the fluxes of matter that are exchanged between the internal components of the Earth system. The former, therefore, has almost no influence on the latter. Of course, the exceptional impact of a large meteorite occurs from time to time with important consequences for evolutionary dynamics, the most well known of them being the extinction of the dinosaurs, with the exception of the ancestors of birds, 65 million years ago. However, such an event, even if it heavily impacts the nature of present and future species, has no long-term effect on the Earth system's general biogeochemical dynamics. Once the event and its climatic consequences have passed, the system regains the normal course of its regulations and evolutions, which we will describe in this chapter and which are also addressed in Chapter 4, and in Chapter 2 of [MON 14b].

If the Earth has had an exchange with the universe in the pst 4 billion years, it is likely to have been through a flux in radiative energy. For thermodynamics, this means a non-isolated, but "closed", system. The Earth receives a certain quantity of radiative energy from the Sun and from space and sends an equal quantity back into space, but increases the average wavelength of its radiation; that is to say it shifts its spectrum to infrared. The individuality of the Earth system is thus partly due to the fact that it is a closed system and its perimeter is not arbitrary since it corresponds to the pragmatic image that we have of it as an object, well separated from the exterior universe. Moreover, we will show later that the maintenance of an atmosphere, oceans, a climate and life, for 4 billion years, is largely due to the fact that the Earth is a closed system through which energy travels and is transformed. This is an exceptional situation in the solar system, since we can observe other neighboring planets, formed at the same time, such as Mars and Venus, where these characteristics are not present simultaneously.

The atmosphere of Mars is extremely thin and liquid oceans do not currently exist on Mars nor Venus. The condition of a "closed planetary system" is therefore necessary, but not sufficient, to explain the preservation of the principal characteristics of Earth – atmosphere, liquid water, life – over the course of 4 billion years. The other conditions stem from adaptive interactions that have been established progressively, on the one hand between the larger components of the Earth system (e.g. geophysical, geochemical and biogeochemical exchanges) and on the other hand between those components and life forms that constitute them (flux due to the metabolisms of life forms). The objective of this chapter is to describe these interactions.

1.1.4. *The major components of the Earth system*

The general physics and biogeochemistry of the Earth system show that the most important internal fluxes of matter and energy only occur within a few particular interfaces. The interface that separates the "solid" Earth from external envelopes of fluids (the ocean and the atmosphere) is found at the location of volcanic emissions, erosion, sedimentation and the subduction of tectonic plates. The interface between the ocean and the atmosphere gives rise to evaporation, precipitation and exchanges of gas. The interface between the continents and oceans is the location of transfer of continental matter eroded into the oceans, in particulate and dissolved forms, that partly fuels the sedimentation of the oceans.

A simplified but pertinent description of the Earth system therefore consists of defining the internal "base" components through a few very large compartments (or reservoirs), which include the geosphere, ocean and atmosphere. They are like the vital organs of an animal between which fluxes of matter and energy travel. In this sense, they are physical compartments, whose role in the Earth system is "organic".

Improvements in our knowledge of these compartments, through active research over the last few decades, have, of course, led to the refinement of the understanding of the system. It is in this way that we are able to make distinctions between the different reservoirs of water on earth (the hydrosphere) – the oceans, continental fresh water, ice and water vapor. However, further distinctions between subcompartments have proven to be necessary. For example, the ocean's surface and its interior do not have the same physical properties and are separated by a discontinuity in temperature (the thermocline) but also a discontinuity in density. These two subcompartments exchange matter and energy between themselves (via thermohaline oceanic convection and turbulence in the thermocline), but also experience exchanges with the atmosphere at its upper surface, in the case of the ocean surface, and with the seafloor, in the case of ocean's interior (underwater volcanic activity, sedimentation, early diagenesis, circulation of hydrothermal fluids, etc.). Similar distinctions exist between the upper and lower atmosphere, or between the geosphere in contact with the atmosphere (soils, outcrops, lavas and basalts on continents and islands) and that which is in contact with the ocean (sediments on continental margins and abysses, lavas and basalts).

1.1.5. *What is the biosphere?*

The notion of a "biosphere", in parallel with the large components that we have just described above, brings us back to considering the sum of all life forms as simply an additional component. Evidently, this would not correspond to the reality and would moreover be fundamentally dismissive in regards to the importance and complexity of the role of life in the functioning and evolution of our planetary system.

Firstly, life forms are neither homogenous nor continuous. They have effectively colonized a great number of niches, whether that is the surface of the geosphere (sediment), the oceans, continental water or the surfaces of continents. Moreover, although they represent only a slight mass (biomass) compared to that of the large compartments of the Earth system, life forms are the mediators of an important part of the flux of matter between compartments through the phenomena of photosynthesis and respiration, the mobility of organisms, the phenomenon of bio-mineralization leading to the construction of internal skeletons and shells, and the production of organic and mineral waste (senescence, death, production of feces and urine, etc.). A life form is therefore not, strictly speaking, a reservoir of the Earth system, but rather a motor, or mediator, of its internal interactions. But the definition of the biosphere does not stop there. When we speak of the evolution and regulation of the Earth system across large scales of time, life is also characterized by its endlessly renewed diversity, from which its adaptive plasticity develops under the pressure of natural selection. It is this plasticity of life forms that, via its role in the exchange of matter, allows the entirety of the Earth system to be evolutionarily flexible.

Thus, we acknowledge that the description of an ecosystem, whatever its perimeter, does not just come down to its life forms. It cannot limit itself to describing instantaneous states, but should also describe the dynamics and retroactions, where living and non-living entities enter simultaneously into play. We will, therefore, conclude here that the term “biosphere” is equivalent to that of the Earth system (or global planetary ecosystem) from the advent of early life on Earth (around 4 billion years ago) to today, as such, for the greater part of its history. But in order to avoid any confusion, we will restrict ourselves to using a single expression: “Earth system”.

1.2. The ocean, from its origins

1.2.1. *Was there an ocean 4.4 billion years ago?*

Isotopic studies carried out on zircons from Australia dating from 4.4 billion years ago [WIL 01] have given rise to the emergence of the hypothesis that liquid water could have been present on the primitive Earth from the end of its principal accretion phase, characterized by the presence of a magma ocean resulting from the transformation of gravitational energy into heat. The liquid water would have accumulated on a solid crust that had only just cooled.

Even if this hypothesis still remains widely disputed, it is generally acknowledged that the oceans were probably formed more than 4 billion years ago. Evidently, this early liquid water was in equilibrium with an atmosphere much denser than that of today, of which the pressure at ground level was several tens of times greater than that of today. The original oceans therefore probably experienced a very hot phase, in the order of several hundreds of degrees Celsius.

1.2.2. *The origin of water on Earth (4.5 – 4 billion years ago)*

It is self-evident to say that water on Earth has an extraterrestrial origin because this is true of all the elements that constitute our planet.

Nothing is created from nothing in our observable universe! Nevertheless, the arrival of water, carried on meteorites and comets, was not in exact synchrony with the arrival of rocky elements and this is due to the history of the formation of the solar system. Fervent research activity carried out by planetologists and astrophysicists particularly endeavors to explain the differences in composition and orbit of objects in the current solar system (the Sun, telluric "rocky", planets, giant "gas" planets, satellites, asteroid belts, comets, the Oort cloud). The explanatory model develops particularly in the form of a historical sequence of several important events such as the formation, differentiation and the progressive cleaning of the protostellar disk, the movement of volatile components (including water) associated with the thermonuclear ignition of the Sun, the formation of telluric planets, the formation of giant gas planets, the migration of Jupiter (the "hot Jupiter" hypothesis) and substantial impacts.

It is not possible to give here an overview of the state of current knowledge in this field; the interested reader can refer to more specialized works on the formation of the solar system. We can, however, emphasize a major point for this present chapter: most water, brought through influx asteroids and comets, accumulated late, after the principal phase of accretion by rocky material, but probably in the first 500 million years of the Earth's history. It is this water that would constitute the majority of the current oceans (93% of water on earth) and the different reservoirs of fresh water (glaciers and ice caps above all, then lakes, rivers, groundwater, atmospheric water vapor and clouds).

1.2.3. *The ocean and the end of the "Venus" phase of the Earth's history (between 4.5 and 4 billion years ago)*

During accretion, the heaviest elements were concentrated by the effect of gravity at the center of the Earth's sphere (especially iron and nickel, but also other, rarer heavy elements), whereas the lightest elements became concentrated on its surface. Taking into account the initial conditions in temperature and radiation, the Earth's very first atmosphere was probably dominated by gaseous and stable components, formed by the chemical combination of light elements such as hydrogen, carbon, oxygen and nitrogen – the most likely candidates being carbon dioxide (CO_2), dinitrogen (N_2) and water (H_2O). This is the reason why, in the absence of natural archives from the era, the current theoretical models propose an initial atmosphere that was very dense, dominated, apart from water, by around 98% CO_2 and 2% N_2; these proportions can vary slightly depending on the models used [GAR 06]. Certain minor or trace components (such as gaseous SiO_2) are not excluded from these hypotheses, particularly because of the very turbulent conditions caused by enormous, repeated meteoric impacts or by the mega-impact that gave rise to the formation of the dual Earth–Moon system.

The primordial atmosphere was probably very dense (from 60 to 160 times the current atmospheric pressure) because it contained, in the form of gaseous CO_2, the carbon that is present today in the form of solids (carbonated rocks and sediments and organic matter, notably fossils). In such conditions, the greenhouse effect was enormous and it kept the Earth's surface temperature at several hundred degrees Celsius. This phase is sometimes called the Earth's "Venus" phase by analogy with the conditions that prevail today on Venus.

Water, which arrived on Earth progressively carried by meteorites and comets, progressively accumulated, at first in gaseous form, increasing the greenhouse effect, then in liquid form when the saturation pressure was reached. It is therefore very probable that the first reservoirs of liquid water contributed considerably to the reduction of the greenhouse effect [LIU 04] as they constituted chemical reactors by which gaseous carbon in the form of CO_2 could be transformed into solid carbon in the form of carbonates ($Metal^{2+}CO_3^{2-}$), via the equilibriums between the different forms of carbon mineral dissolved that coexist in the aqueous phase, CO_2, HCO_3^- and CO_3^{2-} [ZEE 01].

It is helpful to note that this general process is constantly occurring today, but through biologically mediated processes (biomineralization of calcium shells and skeletons) and via the alteration of aluminosilicate rocks. This alteration consumes CO_2 and some of this carbon is ultimately found in the ocean in the form of carbonate ions (CO_3^{2-}) due to oversaturation. These will then be transformed into solid form by the biological processes that create shells and skeletons, then stored in the sedimentary reservoir. Details on these processes can be found in Chapter 4, and in Chapter 2 of [MON 14b].

It is unknown how much time the end of this "Venus" phase could have taken; many specialists on the primitive Earth now believe that conditions favorable to the emergence of life were already present 4 billion years ago, that is to say, before the late meteoric bombardment, and it is not impossible that an emergence of life may have taken place in the ocean from this era [GAR 06]. If it existed, did this life survive a late meteoric bombardment? Did other emergences, successful or abortive, take place later? These questions remain largely open.

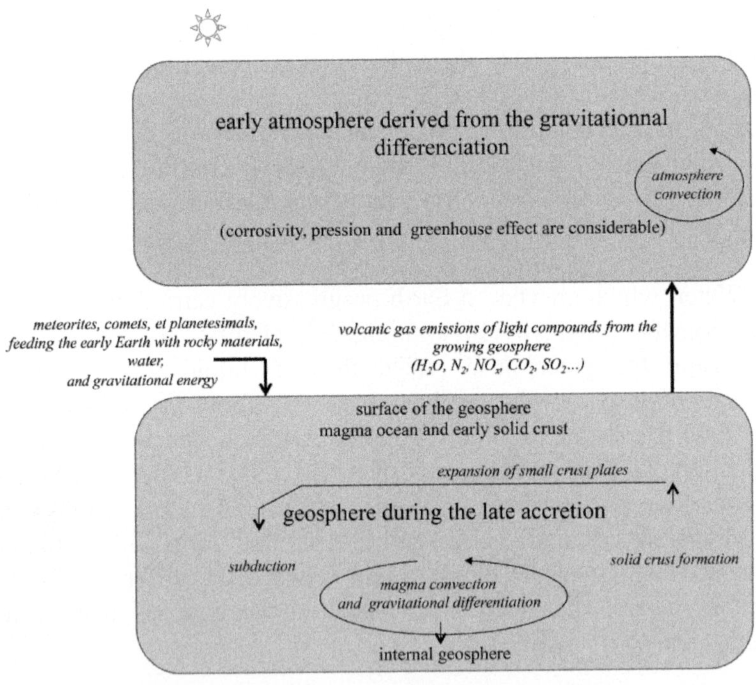

Figure 1.1. *Diagram of the Earth system at the end of accretion around 4.5 billion years ago (see color section)*

COMMENTARY ON FIGURE 1.1.– The influx of evaporation into the atmosphere is not compensated for and the planet is surrounded by a very dense atmosphere mainly composed of carbon dioxide (CO_2), dinitrogen (N_2) and gaseous water, resulting from gravitational differentiation (the heaviest elements being concentrated as a core at the heart of the geosphere). The greenhouse effect is enormous and an ocean of water cannot form. This is the Earth's "Venus" phase.

1.2.4. *Why are there oceans on Earth and a "Venus inferno" on Venus?*

Earth and Venus are two neighboring telluric planets, formed at the same time and of a very comparable size. How do we explain the fact that the conditions on their surfaces and in their atmospheres are so dissimilar today? [1]

Aside from the conditions on their surfaces and the composition of their atmospheres, two important characteristics distinguish the two planets. First, there are no active plate tectonics on Venus, whereas Earth maintains one. We can observe this in everyday life through earthquakes, volcanic activity or, more poetically, through contemplating our mountainous countryside. The second characteristic is the presence of an important satellite, the Moon, which revolves around the Earth. The Moon and the Earth have a very important mass ratio (1/81), now known to us as the most important between two planetary bodies in orbit around their common center of gravity (which is naturally much closer to the center of the Earth than to the center of the Moon). This duo resulted from a giant impact between two planets in the process of formation just over 4.5 billion years ago. Because of its kinetic conditions (notably speed and angle of incidence), this impact threw out a very significant quantity of matter into space while still preserving the core of the larger of the two

1 Remember the story that Venus, the morning star, was known to the Romans by the name of Lucifer (carrier of light), whereas the Christian tradition ended by transposing Lucifer, the angel dethroned after wanting to supplant his creator, into Satan.

proto-planets. The result was the formation of an orbital disk, followed by the progressive cleaning of this disk to form the Moon [KOK 00].

However, another scenario could have involved the instantaneous introduction of a significant quantity of water into the magma mantle of this protoplanet, a precursor to Earth in the right conditions, an event that did not occur, or which occurred differently, in the formation of Venus [BIB 09]. As we now know, thanks to the study of plate tectonics on Earth, the filtration of light components (including water) into the mantle leads to a decrease in its viscosity [BIL 01, DIX 04] and to a chemical differentiation between the continental crust and the ocean along the subduction zones. Together, these two phenomena favor the maintenance of an active tectonic system, at least as we know it on Earth.

Some results of the exploratory mission *Venus Express* published in 2009 by the European Space Agency [ESA 09] suggest that a tectonic system nonetheless existed at the beginning of Venus' history, but it would have ceased to develop due to the lack of ocean. The primordial water was in fact systematically vaporized with a positive retroactive effect on the greenhouse effect, leading to its enhancement and to the gravitational escape of hydrogen.

In the case of the Earth, one hypothesis is that a much more active tectonic system, initiated by the giant impact that led to the formation of the Earth–Moon duo [RUI 11], could have absorbed some of the primordial water via subduction, sufficiently slowing the enhancement of the greenhouse effect so that sizeable oceans could form. After that, the reduction of the initial greenhouse effect by the precipitation of carbonates could have become considerable and the first stage of abiotic regulations of the Earth system was established. This is illustrated in Figure 1.2.

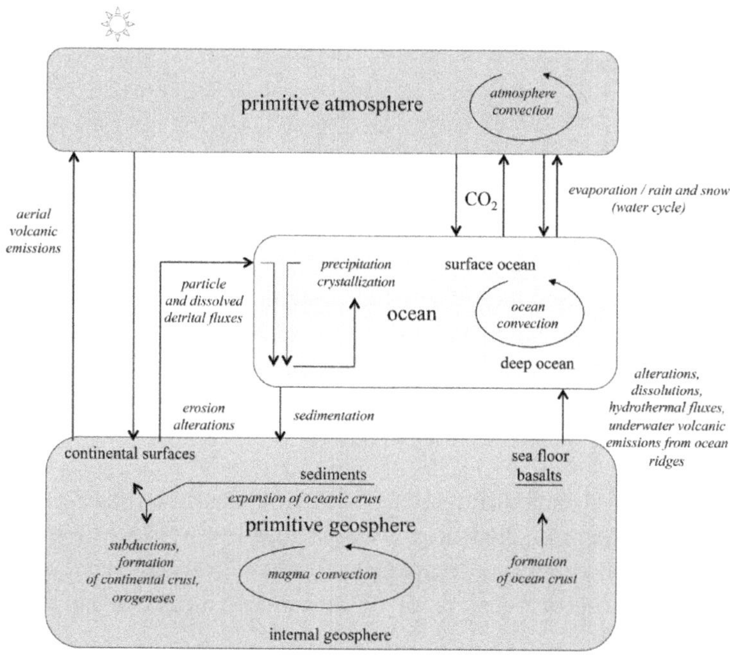

Figure 1.2. *Diagram illustrating the bases of the first abiotic regulations of the Earth system (see color section)*

COMMENTARY ON FIGURE 1.2.– Due to an ocean being in place, the inward fluxes are from now on potentially compensated by outward fluxes for each of the reservoirs of matter. Short-term regulation is assured by the chemical equilibriums of the ocean (the detail of which will be addressed later), whereas the long-term regulation depends on the tectonic cycle. The tectonic activity of the Earth system has been maintained up until today, whereas that of Venus ceased to be very early on. It is probable that this difference arises from the permanent introduction of water into the mantle via the phenomena of subduction. Such a communication between the upper geosphere and a liquid ocean would be established on the Earth system while the early tectonic was still sufficiently active, perhaps because of a more significant initial hydration of the mantle resulting from the mega-impact that formed the Earth–Moon duo.

1.2.5. *The ocean, cradle of the first living creatures (between 4.4 and 3.5 billion years ago)*

What criteria can we use to distinguish a living state from a non-living state? This question is an inexhaustible source of impassioned conversations where different viewpoints clash, from those of scientists (the presence of a cellular metabolism, the presence of a genetic "code", the presence of a capacity for "reproduction", the presence of an open, self-regulated system, the presence of a steady state maintaining or increasing the level of organization, that is to say locally going against the second principle of thermodynamics, etc.), up to those beliefs of a spiritual, philosophical or religious order (essentialism, vitalism, creationism, etc.).

On a strictly scientific level, does this question have only one meaning? Is there a scientifically based reason for which it would be necessary to distinguish a living state from a non-living state, or is this only, once more, a question of definition and limits of an arbitrary nature?

To develop an argument in reply to this question here would take much too long and would not be relevant to this work. This is why we will restrict ourselves here to defining, arbitrarily, the "minimal" state of a life form as that of a biological cell possessing a metabolism, genetic information and a capacity for reproduction. This arbitrary definition should be distinguished from that of life in a general sense, which is richer and more complex, and to which we will return later.

A biological cell needs water, and the chemical prebiotic systems that preceded the first cells also needed it. The presence of liquid water, indeed of an aquatic environment, is therefore a necessary condition for the emergence of life. Researchers working on prebiotic conditions and the first life forms do not rule out the possibility that these could have appeared in multiple locations and on multiple occasions, in isolated lakes or seas that probably coexisted and succeeded each other in different places and at different stages of the initial evolution of our planet, even 4 billion years ago [GAR 06]. However, the geological evolution (primitive tectonics), the environmental variability (local climate, deposits of sediment and drought) and major hazards (volcanic activity, falling asteroids, notably before and during the last phase of meteoric bombardment around 4 billion years ago) render the survival of small, local autonomous ecosystems on timescales in the order of a million years or more almost impossible.

The question of the "origin" of life in the Earth system is therefore not only that of the processes by which the first living beings appeared in general, but it is also that of the conditions that permitted one of these life forms to become the common ancestor of all life today. How has life been able to develop into the lattest form, i.e. human beings, over billions of years, as phylogenetic trees indicate, which stretch back to a last common ancestor having lived 2 or 3 billion years ago (*Last Universal Common Ancestor* (LUCA)) and from the findings of paleontology, which stretch back to around 3.5 billion years ago (the age of the most ancient known bacterial fossils, in the form of stromatolites)?

The presence of a sufficiently developed ocean at different latitudes and in different climates is an element of enquiry that concerns the crucial phase of development of the first ecosystems. For life to have been able to "survive" hazards of all types in the long-term, it is necessary that its random destruction was only partial and rapidly compensated by a new colonization of the destroyed environments. It is necessary to underline here that the colonization is not a response to the necessity of increasing the chances of survival of the entirety of the population, but simply that of increasing the chances of survival of each individual by finding the resources that it needed elsewhere. The necessity is individual, but its effects extend to the scale of populations and ecosystems.

There is therefore every reason to think that the life from which human beings originated appeared and developed in an already significant ocean. That this origin may have been associated with deep hydrothermal environments through the use of chemical energy available at high temperature [REY 01, SHO 96], or rather in the surface environments through the use of solar energy, is a question that remains very open to research. Many of the life forms from deep hydrothermal environments, notably animals, have adapted to these extreme nutritional niches by natural selection, beginning from the ancestors that they shared with bacteria, cephalopods, fish, or crustaceans from other marine environments. However, it is unknown whether it is the same for thermophile and chemosynthetic bacteria or if these forms of life appeared locally.

1.3. The ocean, oxygen and the evolution of life forms

1.3.1. *The essential characteristics had been selected in the ocean before the Cambrian period, over 540 million years ago*

The work of the famous American paleontologist Charles Doolittle Walcott led him to discover, at the beginning of the 20th Century, the extraordinary diversity of fossils in the "Burgess Shale" (a fossil field in British Columbia, Canada) [WHI 80]. The corresponding organisms, dating from the Cambrian period (between 540 and 500 million years ago – even if the majority of fossils are in the *later* Cambrian period), were the most ancient forms of animal known at the time. They were marine animals, characterized by the presence of endoskeletons or exoskeletons and by a body organized into segments, such as we can still find today in arthropods, insects and their distant cousins: the vertebrates. This discovery formed the origin of the idea that an accelerated diversification of living species, known as "the Cambrian explosion", occurred at this time. The exceptional preservation of these fossils includes not only their carapaces (toughened external body parts, e.g. an exoskeleton or shell), but also morphological traces of soft tissue situated in the area of the body. It was therefore legitimate to delve deep in the study of the diversification, functioning and behavior of these organisms.

In light of today's knowledge, it is sensible to acknowledge this idea of an evolutionary explosion belonging to this era and to which we can associate the genesis of the entirety of living things. After close inspection, the fossilized organisms of the Cambrian era are for the most part animals; living beings already displaying certain fundamental characteristics in common, selected for over the course of numerous former diversifications. Some of them (on which there will be a new question later) are aerobic organisms: symbiotic cells with a differentiated nucleus (eukaryote) and a characteristic multicellular organisation. Furthermore, the development of carapaces is an evolutionary innovation of the Cambrian period which naturally favored the preservation of fossil traces of animals in possession of these attributes, and which focused the attention of paleontologists on the Cambrian and post-Cambrian periods. Indeed, we know today that many more ancient marine animals without carapaces existed 600 million years ago (the Ediacarian fauna, after the name of the Australian site where these fossils were discovered). This is evidence of the diversification of multicellular organisms in this era, plants as much as animals as no fauna is possible without flora. Some proterozoic fossils dated from 2.1 billion years ago have been found recently in Gabon [ELA 10]. Their morphological complexity intrigues paleontologists and leads to the belief that they could be either multicellular organisms (plant and/or animal) or colonies of prokaryotic cells (bacteria or archaea) possibly in possession of a rudimentary form of functional organization, highlighted by the work of microbiologists using modern examples [KEI 04]. Whatever they are, numerous important diversifications have occurred since the origin of terrestrial life forms, before the Cambrian period, to promote those that led to the selection of the molecular intracellular system that we know today (DNA/RNA/proteins) and of all its genomic and functional variants. The largest number of these diversifications occur in the domains of bacteria and archaea: unicellular prokaryotic microorganisms (cells without a differentiated nucleus), in contrast to eukaryotic cells whose diversity remains largely unexplored.

Figure 1.3 illustrates the principal evolutionary innovations that occurred, mainly in the ocean, between the beginning of life on Earth (over 3.5 billion years ago) and 400 million years ago (early Devonian Period). It also shows the major environmental evolutions, from an exclusively anoxic environment to a very heavily oxygenated environment, where only a few anoxic niches remained (organic sediments, digestive tubes and poorly ventilated deep zones of the ocean or certain lakes). These environmental evolutions matched evolutionary developments, such as the emergence of oxygenic photosynthesis, at the same time creating a new framework of parameters for later developments, such as the emergence of aerobic respiration.

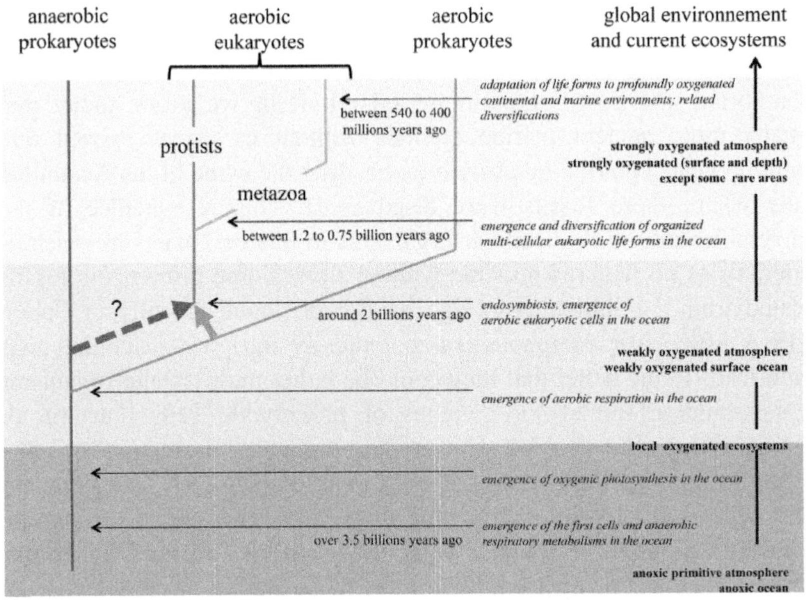

Figure 1.3. *The main evolutionary developments and the major environmental evolutions between the beginnings of life on Earth (over 3.5 billion years ago) and 400 million years ago (early Devonian Period) (see color section)*

The links of cause and effect between the oxygenation of the environment and the major developments previous to the emergence of aerobic respiration are however less evident and still constitute working hypotheses for research. In the emergence of eukaryotic cells, it is not unreasonable to assume that prokaryotic cells, subjected to an increasingly oxygenated environment, were able to find mutual benefits by sharing respective innovations in the framework of endosymbiosis [MAR 91]; some taking the best part of their energy from aerobic respiration (mitochondria) and others furnishing the protective mechanisms of cytoplasm or DNA against the toxic effects of oxygen. As far as the emergence of multicellular organization is concerned, it is not unreasonable to assume that such an organization could have caused external contact with oxygen to be reduced. The aim of natural selection would then be to specialize the external cells in resisting the toxic effects of oxygen and to ensure an influx of matter into the internal cells. A major question for research is to discover how the change occurred from a malleable and temporary adaptation, responding to external conditions, to a biologically cemented situation, where it is the endosymbiotic cell or colonial cell group that reproduces itself as such.

Finally, the colonization of continental environments can also be linked to the history of oxygen. An atmosphere globally enriched with O_2 is favorable to the formation of ozone (O_3) because of the photochemical reactions that occur in its upper layers, which are subjected to very powerful rays from the Sun. The formation of a permanent ozone layer, even if subject to fluctuations with the seasons, certainly favored, indeed rendered possible, the selective adaptation of marine organisms in continental environments. The term "out of water" is too often used incorrectly here, since the organisms adapted first to environments from which water periodically retreated, such as in the intertidal zones of marshes, up to the point where they were capable of leaving the water for increasingly longer periods.

On a global scale, life forms and their environments have thus coevolved, through the device of retroactions, for over 3.5 billion years. Upon careful observation, we can confirm that the most fundamental characteristics of present day life forms (aerobic or anaerobic respiratory metabolism, endosymbiosis of eukaryotic cells and multicellularity) were all selected for in the ocean, before the start of the Cambrian period, over 540 million years ago.

1.3.2. *How did oxygen accumulate?*

Oxygen is an element very receptive to electrons; is it therefore very chemically reactive. The word "oxidize" comes moreover from this property since oxygen, in its O_2 form, is capable of oxidizing almost all the other elements or chemical compounds. Only a few halogens, such as fluorine or chlorine, hyperoxygenated anions (permanganate MnO_4^-, dichromate $Cr_2O_7^{2-}$) or ozone (O_3), have a greater oxidizing power.

Because of this property, the gaseous form of oxygen, the form that nevertheless composes 21% of our current atmosphere, is extremely unstable. In this form, oxygen has a tendency to react with the majority of bodies (gas, liquids or solids) with which it comes into contact, to form oxides while releasing energy. Combustion is an accelerated manifestation of oxidation, maintained by the large heat flux it gives off (see also Chapter 4 of this book, and Chapter 2 of [MON 14b].

Consequently, the massive presence of O_2 in our environment can only be explained by a shortage of potential chemical partners and/or by a permanent production of O_2 stimulated by an energy supply. As we will describe, these two conditions have almost always coexisted in the history of the Earth, even if the availability and nature of the chemical partners have varied in the course of time. If the atmosphere and, later, the continental surfaces, have been important players in this history, it is in the ocean that it essentially occurred.

The Earth's primitive environment was strictly anoxic, i.e. O_2-free [HOL 84, SCH 10]. Even supposing that some rare abiotic processes were locally susceptible to form O_2, this would immediately consume the poorly oxidized (reduced) compounds that surrounded it and would immediately disappear, in this process of chemical "predation".

Starting from the first living cells, the most efficient metabolic mechanisms, at first anoxic, were selected to supply to these cells the energy and matter that they needed. Some of them were anaerobic mechanisms for photosynthesis. They allowed the cell to use, at least partially, solar energy to carry out the synthesis of organic molecules necessary for maintaining cellular machinery just as for its development and duplication. The cells in possession of this new mechanism gained the opportunity to free themselves from the need to capture and digest, through respiration or anaerobic fermentation, other cells or organic molecules available in their environment (heterotrophy). This evolutionary development was likely selected for because the energy supply of light from the Sun was less random than that of organic fuel in the first ocean. Some anaerobic photosynthetic cells still exist today in a few rare anoxic environments where light manages to penetrate. They live, for example, in the subsurface of certain fine deposited sediments subject to marsh tides [SEI 93].

To simplify the discussion, below, we will only speak about oxygenic photosynthesis (which produces oxygen) and leave aside the more anecdotal case of photosynthesis which does not produce oxygen, even though it could have played an important role at the very beginning of life on Earth.

The simplified equation for oxygenic photosynthesis is:

$$H_2O + CO_2 + \text{light} \rightarrow CH_2O + O_2 \qquad [1.1]$$

This formula is the same for photosynthetic anaerobic life forms (which can only live in the absence, or near absence, of oxygen, essentially bacteria) and aerobic life forms (bacteria, algae, terrestrial plant life). In effect, oxygen is certainly a product of this formula, but this does not imply that O_2 is locally concentrated to the point where photosynthetic cells are necessarily aerobic. The adaptation of these cells, aerobic or anaerobic, can differ according to the confinement and the concentration of oxygen that results, but the balance in carbon and oxygen from oxygenic photosynthesis is always the same.

The CH_2O is evidently only an approximation of the average proportion of carbon, hydrogen and oxygen of the organic molecules of the cell. Moreover, this formula neglects the role of the other necessary elements (nitrogen, phosphorus, iron, etc.).

The key trait of this balance is that the production of organic matter by photosynthesis is accompanied by a production of oxygen (O_2). The anaerobic photosynthesis in the ocean was the first source of O_2 in the terrestrial environment since the first ocean was an anoxic environment. Still, the existence of this source was certainly a necessary, though not sufficient, condition for the oxygenation of the environment. This condition, still true today, can only be produced by the ocean, even if the continents are now largely colonized by plants. This condition is explained in the paragraph below.

The oxygen produced by photosynthesis is, as we have already seen, highly unstable. It therefore reacts immediately with numerous compounds or chemical elements present in the primitive anoxic environment. It reacts, for example, with sulfurs (S^{2-}) to produce sulfates (SO_4^{2-}). We note in passing that it thus recycles the oxidizing compounds necessary for certain anaerobic (sulfate-reducing) cells to fulfil their energy needs, since these "respire" sulfates by transforming them into sulfurs. In consequence, without the oxygen issuing from anaerobic photosynthesis, any aerobic respiration would not have been able to maintain itself or to develop in the long-term in the terrestrial environment. A massive and non-localized source of external energy was necessary, which only solar rays could furnish via photosynthesis.

However, O_2 also reacts with CH_2O to form the more stably composed CO_2, releasing a small amount of heat as it does so. This is expressed by a formula strictly inverse to photosynthesis, where only the form of the energy is modified but not its quantity. The following formula summarizes the oxidation of organic matter:

$$CH_2O + O_2 \rightarrow H_2O + CO_2 + \text{heat} \quad [1.2]$$

For an excess of O_2 to accumulate in the environment, it is thus necessary that the availability of its chemical partners, (whether sulfur or organic matter, but also oxidizable ions in solution such as Fe^{2+} (which is transformed into solid iron oxide Fe_2O_3)), is inferior to the flux of oxygen produced by photosynthesis.

The precipitation of solid oxides did, certainly, consume some of the chemical prey present in the ancient oceans. We can observe traces of it within certain sedimentary accumulations of the Archean Period, for example in the banded iron formations at Barberton in South Africa dated around 3.2 to 3.5 billion years ago [HOF 05]. Nevertheless, the key to the accumulation and maintenance in the long-term of an excess of O_2 in the environment, up until to today, is found in the fact that a proportion of the organic matter produced by photosynthesis is constantly transferred into oceanic sediments and, thus, prevented from oxidation by O_2.

$$H_2O + CO_2 + \text{light} \rightarrow \text{sedimentary } CH_2O + O_2 \text{ in the environment} \quad [1.3]$$

Even if this sedimentary flux of organic matter is very small relative to the flux exchanged during photosynthesis and the oxidation of organic matter, its accumulation over long periods of time has enabled the construction of important stocks. The O_2 oxygen in our environment can thus be considered as the counterpart of the global stock of organic sedimentary matter, the living biomass being quantitatively negligible in relation to it.

1.3.3. *The first important accumulation of oxygen (around 2.5 billion years ago)*

Although sedimentary archives stretching back several billion years are rare and fragmented, it has been possible to show that a first important global accumulation of oxygen was produced around 2.5 billion years ago [CAT 05, VOE 10]. This accumulation increased the partial pressure of O_2 (pO_2) at ground level from around 10^{-5} bar to $10^{-1.5}$ (= 0.03) bar [CAT 05], in equilibrium with the oxygen dissolved in the ocean's surface. This first oxygenation only represented around 15% of its total transformation into the pO_2, of which the current atmosphere is 0.21 bar. It nevertheless considerably modified the course of biological evolution just as that of the evolution of the Earth system.

It is not possible to confirm if this first accumulation was principally linked to aerobic or anaerobic photosynthesis, or both. A local and temporary oxygenation of the medium is undoubtedly a prerequisite produced where, in a generally anoxic world, photosynthetic marine ecosystems were the most productive (in certain seasons). Such fluctuations created toxic environments for the photosynthetic cells, since they were initially anaerobic; they poisoned themselves.

We cannot dismiss that aerobic photosynthesis, the type which principally occurs today, could have appeared very quickly after anaerobic photosynthesis and that it could have rapidly become the principal contributor to the global accumulation of oxygen.

1.3.4. *A moderate increase in oxygenation (between 2.5 and 0.5 billion years ago)*

After the first rapid accumulation of oxygen, around 2.5 billion years ago (beginning of the Proterozoic Period), it is very likely that the terrestrial environment continued to accumulate oxygen, albeit more slowly. Reconstructions [BER 04] suggest levels of O_2 in the atmosphere were in the order of only 10 – 15% 500 million years ago (either 0.1 and 0.15 bar of pO_2 for a hypothesis with an atmospheric pressure of 1 bar). This signifies a very slow augmentation average, in the order of 4% volume increase per billion years.

The cause of the deceleration that followed the initial rapid accumulation of O_2 is based on evolutionary adaptations that permitted an increase of oxygen recycling from organic matter, thus consuming more of the oxygen produced by photosynthesis. The first of these adaptations was aerobic respiration, which is only, in fact, the biological version of the oxidation of organic matter by oxygen as described above. This adaptation could only appear after anaerobic photosynthesis, under the selective pressure of an environment at least temporarily oxygenated, and could only spread with the first global accumulation of oxygen.

Aerobic respiration has a better energy yield than anaerobic respiration. As such, it provides the cell the advantage of counteracting the toxic effects of oxygen. In this case, it is possible that the preferential oxidation of certain molecules (external fuel or synthesized reserves of the ATP type) enabled the preservation of essential molecules of the cellular machinery (DNA, RNA, proteins).

From the moment aerobic respiration was selected for, other evolutionary refinements could develop, particularly in the ocean.

This was first the case of endosymbiosis (an aerobic eukaryotic cell issuing from the symbiosis of several initially prokaryotic cells), just over 2 billion years ago, followed by that of the emergence (in several separate stages), of the first organized multicellular life forms: fauna and flora consisting of eukaryotic aerobic cells [RUI 07, SAC 08]. Multicellular life forms were already present in the Ediacarian Period (fauna from around 600 million years ago) and phylogeneticists generally place their appearance between 1.2 and 0.75 billion years ago. It is, however, possible that they are older. This is suggested by a few rare paleontological discoveries, but whose interpretations are still much disputed [ELA 10, RAS 08].

All these evolutionary refinements contributed to the improvement of recycling organic matter and limited the global accumulation of oxygen in the environment. At the same time, evolution prepared itself for a new acceleration of global oxygenation, which we will now describe.

1.3.5. *The second important accumulation of oxygen (between 500 and 350 million years ago)*

As of 500 million years ago (end of the Cambrian Period), two major characteristics were already in place. They constituted the contingence that would make possible the colonization of continental surfaces by life forms.

The first characteristic was a large atmospheric O_2 content in the order of 10 – 15% so that a stratospheric ozone (O_3) layer could form, which filtered the most harmful of the Sun's rays. It is probable that these conditions were established before or from the start of Cambrian Period, over 500 million years ago.

The second characteristic was the evolutionary diversification of the forms and functions of multicellular marine life forms, such as those attested by the Cambrian fauna of the Burgess Shale [WHI 80], which date back to 500 million years (see section 1.3.1). The development of limbs for pinching or moving, and equally the presence of carapaces, make the progressive adaptation of animals to intermediate environments, where water periodically retreated, possible. Some of these organisms probably developed the capacity to briefly leave the water, then to move into an aerial medium, after which point they would rejoin the water by their own means. In this way, they perhaps promoted the adaptation of marine bacteria to continental sunlight, notably those N_2 nitrogen-fixing bacteria that terrestrial plant life needed. Unlike their marine equivalents, the terrestrial plants are only slightly directly provided with nitrate and they depend largely on the fixation of nitrogen (and transformation into nitrate) for their photosynthesis. We note, however, that even in the ocean we can find situations where nitrate is lacking, which leads to the use of N_2 through this fixation.

The colonization of continental surfaces by life forms consequently gave rise to new adaptations, but was also the source of an increased influx of organic matter into the sedimentary reservoir of the Earth. Rivers drained organic debris, essentially plant matter, from the watersheds into the sediments of flood plains and deltas. On the other hand, from the Devonian Period (around 400 million years ago), the evolution of flora and the geographical characteristics of the continents enabled the accumulation of large quantities of peat and coal, the maximum flux having been attained in the Carboniferous Period (around 350–300 million years ago).

As explained above, the sedimentary burial of organic matter triggered a global accumulation of oxygen in the environment and it is precisely between 400 and 350 million years ago that a new acceleration in oxygenation occurred [BER 04, CAN 07]. It may have increased the atmospheric O_2 level to around 30–35%, the most the Earth system has ever known. However, it is necessary to note that this value is considered too high by some paleoenvironmentalists, since the risk of natural combustion exists at 13%, becoming a substantial risk at 25%. Natural fires, of which we find a permanent trace in sediments from the Devonian Period (carbonaceous particles of the "fusinite" type), definitely limited the maximum oxygenation of the terrestrial environment. Moreover, it is uncertain that the capacity for regulating the internal oxygen pressure of multicellular eukaryotes [MAS 03] could have adapted to a hyperoxygenation of the environment. If it could not, the oxygenation would have been self-limited by the drop in biological productivity of the eukaryotes (notably that of plants).

After having reached its saturation point, the atmospheric oxygen level diminished, probably due to a partial release of sedimentary organic carbon accumulated in the course of the Devonian Period (410–360 million years ago) and the Carboniferous Period (360–295 million years ago). 200 million years ago, the atmospheric oxygen level would fluctuate around a value close to the current one (21% in volume), which leads us to think that the flux of oxygen consumed by the oxidative destruction of recent or sedimentary organic matter, which is principally due to aerobic respiration, became equivalent to that of oxygen produced by photosynthesis. This also means that the flux of organic carbon released from old sediments through erosion, then oxidized, is balanced in the long-term by the flux of photosynthetic organic carbon buried in new sediments. The Earth system reached here its long-term regulation (millions of years), one that simultaneously integrated exchanges of matter between the ocean, atmosphere and sediments, the result of 3 billion years of biological evolution, and the tectonic plates that enabled the recycling of sediments *via* erosion and volcanic activity.

1.4. The regulation of the greenhouse effect by the ocean

1.4.1. *There is no life without a minimum greenhouse effect*

Without CO_2 in the atmosphere, the atmospheric greenhouse effect would be extremely reduced and the average temperature at the surface of the Earth would be –18°C, much lower than it is today (+15°C). Such conditions would have been unfavorable for the presence of life forms on the surface of the Earth and would have limited them to a few rare oases where they would have had every likelihood of being destroyed by natural hazards (asteroid falls, significant volcanic events, etc.). For life to persist for several billion years, it was necessary that the life forms were able to extend their area of distribution very greatly and to diversify, as they had done in the oceans, on the continents. This could not occur without a minimum greenhouse effect, of which the regulation was ensured, and still is, by the ocean and its exchanges with the atmosphere on the one hand and sediments on the other.

1.4.2. *The regulation of the greenhouse effect by the ocean*

Firstly, it is useful to recall that today carbon is very unequally distributed between the atmosphere, ocean and sediments. For one atom of carbon present in the atmosphere (in the form of CO_2), 65 are present in the ocean (mainly in the dissolved forms HCO_3^-, CO_3^{2-} and CO_2) and 150,000 are present in the sediments (mainly in the form of calcium carbonate and organic matter).

On the one hand, the level of CO_2 in the atmosphere results from carbon dioxide exchanges, which tend to permanently establish equilibrium between the partial pressure of CO_2 in the lower atmosphere and the partial pressure of CO_2 dissolved in the surface ocean. On the other hand, the ocean can exchange carbon with the sediments through the precipitation of calcites (carbonates of calcium, $CaCO_3$) from dissolved CO_3^{2-} or, conversely, through them being dissolved.

The exchanges of inorganic carbon between the ocean and atmosphere (exchanges of CO_2 between its dissolved form in the surface ocean and its gas form in the atmosphere) and between the ocean and the sediment (precipitation or dissolving of $CaCO_3$) depend largely on chemical equilibriums which tend to establish themselves between the different forms of inorganic carbon dissolved in the ocean. A unique and simplified expression of these equilibriums is as follows:

$$2HCO_3^- \leftrightarrow CO_3^{2-} + CO_2 + H_2O \qquad [1.4]$$

In the current conditions of weak acidity in the ocean (pH = 8.2), these equilibriums lead to the significant presence of carbon in the form of dissolved bicarbonate (HCO_3^-) and minor proportions in the form of dissolved carbonate (CO_3^{2-}) and CO_2. The readers who are interested by the links between the pH of seawater and this speciation of dissolved inorganic carbon will find further details in Chapter 4, and in Chapter 2 of [MON 14b], and equally in the specialized work of Zeebe and Wolf Gladrow [ZEE 01].

Here, we can immediately grasp that the current greenhouse effect, whose modification is mainly linked to variations in the atmospheric oxygen level of CO_2, cannot significantly diminish in the very stable chemical conditions in the ocean.

All increase in CO_2 in the atmosphere is compensated by a change in equilibrium that tends to increase the major reservoir of HCO_3^-, but to diminish in parallel that of CO_3^{2-} dissolved in the ocean and to a slight degree, acidify the ocean (a drop in pH). The related subsaturation of the ocean in CO_3^{2-} ions tends to be compensated by the dissolution of CO_3^{2-} ions present in the calcite sediments. In parallel, it creates conditions where the bioprecipitation of $CaCO_3$ by life forms is made more difficult, which can damage species with shells and skeletons. These retroactive effects restore the initial situation by having imperceptibly modified the enormous sedimentary reservoir of calcite sediments.

Conversely, all loss of oceanic CO_2 by filtration into the atmosphere tends to be compensated, *via* the equilibriums of inorganic dissolved carbon, by a supersaturation of dissolved CO_3^{2-} and a transfer of dissolved inorganic carbon to solid carbonate, notably *via* the biological path (bioprecipitation and subsequent sedimentation), implying only a slight relative drawing on the reservoir of inorganic carbon dissolved in the ocean (mainly in the form of HCO_3^-).

Here, therefore, are two inverse oceanic mechanisms by which the major short-term regulation (a few centuries, the time oceanic convection takes) of the Earth system's greenhouse effect occurs. The ocean thus plays the role of a real climatic "mediator" which responds to perturbations that can induce other natural (fluctuation in emissions of CO_2 linked to volcanic activity) and anthropogenic phenomena (solid carbon destocking by combustion of organic fossil carbon and industrial calcination of calcites for the production of cement).

In the longer term (thousands to millions of years), certain conditions can lead to a significant change in the greenhouse effect. This was the case for episodes of the "Snowball Earth", similar to the Varanger glaciation [HOF 98] which occurred between 900 and 540 million years ago, that is to say, before the Cambrian Period. Numerous authors consider that such episodes were not total glaciations, but that they preserved the hydrological cycle and the life that is associated with it across a peri-equatorial unfrozen belt that became larger or smaller according to the seasons. These phenomena could have been caused by phases of fragmentation of super continents such as Rodinia, the most ancient of the super continents that would have been formed more than a billion years ago and that would have fragmented 750 million years ago [DON 04]. The volcanic phenomena that would then be produced would have been similar, but much more extended, to those of the current large East African rift and would have greatly extended the basalt surface exposed to chemical weathering, favoring the following sequence of transformations:

– chemical weathering of aluminosilicates (the example of anorthite is used here), which consumes two molecules of atmospheric CO_2 to dissolve each molecule of aluminosilicate, and transport of the dissolved products by rivers:

$$[CaAl_2Si_2O_8]_{solid} + 3H_2O + 2CO_2 \rightarrow [Al_2Si_2O_5(OH)_4]_{solid} + [Ca^{2+}]_{dissolved} + 2[HCO_3^-]_{dissolved} \quad [1.5]$$

– adjustment of the equilibriums of inorganic carbon dissolved in the ocean, restitution of one molecule of CO_2 to the atmosphere and creation of a supersaturation of the ocean in CO_3^{2-}:

$$2[HCO_3^-]_{dissolved} \rightarrow CO_2 + [CO_3^{2-}]_{dissolved} + H_2O \quad [1.6]$$

– precipitation of solid carbonate in the ocean and its stocking in the sedimentary reservoir:

$$[CO_3^{2-}]_{dissolved} + [Ca^{2+}]_{dissolved} \rightarrow [CaCO_3]_{solid} \quad [1.7]$$

The equilibrium of inorganic carbon dissolved in the ocean having been adjusted, the final result of the alteration of aluminosilicates therefore really constitutes a sink for CO_2. If this loss is not rapidly compensated by volcanic emissions, the greenhouse effect can greatly diminish and provoke very widespread and long-lasting glaciations (over millions of years).

Another possibility for significant and long-lasting variations of the greenhouse effect depends on long-term variations in the alkalinity of the ocean.

The paragraph below calls on the notion of the *conservative* chemical form in seawater. It is therefore important first to define this notion. A dissolved chemical form is called *conservative* in seawater if it does not transform into any other in the usual oceanic conditions. For example, this is the case for ions such as Na^+, Cl^- and SO_4^{2-}. In contrast, the molecule CO_2 or the ions HCO_3^- and CO_3^{2-} can transform through the equilibriums summarized in equation [1.4], and are called *non-conservative*.

The pH of the current ocean (pH 8.1– 8.2) is greater than that of pure water (pH 7.0) because the slight electric imbalance (an excess in positive charges) between the conservative cations (essentially Na^+, Ca^{2+}, Mg^{2+} and K^+) and the conservative anions (essentially Cl^- and SO_4^{2-}) is compensated by the slight electric imbalance (excess of negative charges) of non-conservative ions (mainly HCO_3^-, CO_3^{2-}, $B(OH)_4^-$, OH^- and H^+). As the excess of positive charges is mainly due to the alkaline cations, the concentration of charges of non-conservative ions that compensates it is called the total alkalinity of seawater [DIC 81, ZEE 01]. A simplified form of alkalinity for practical use is as follows:

$$AP = [HCO_3^-] + 2[CO_3^{2-}] + [B(OH)_4^-] + [OH^-] - [H^+] \qquad [1.8]$$

The alkalinity of seawater can be significantly modified in the long-term (thousands to millions of years), essentially because of the variation in concentrations of conservative ions. Moreover, all things being equal, the concentrations in non-conservative ions, so also the pH (which reflects the concentration of ions H^+), are only involved in adjusting to the change in alkalinity. Nevertheless, this adjustment can cause a variation in the greenhouse effect. Indeed, the concentrations in H^+ ions (so the pH) and in HCO_3^- and CO_3^{2-} ions, which contribute to alkalinity, are interdependent *via* the equilibriums of dissolved inorganic carbon, as shown in Figure 1.4. If the ocean becomes more acidic (a drop in pH), the levels of CO_2 and HCO_3^- increase while that of CO_3^{2-} diminishes. The consequences are an exportation of CO_2 into the atmosphere, so an amplified greenhouse effect and conditions favorable to the dissolving of solid carbonate. These conditions are detrimental to the existence of some organisms that construct shells or skeletons. The consequences are inversed if the pH increases.

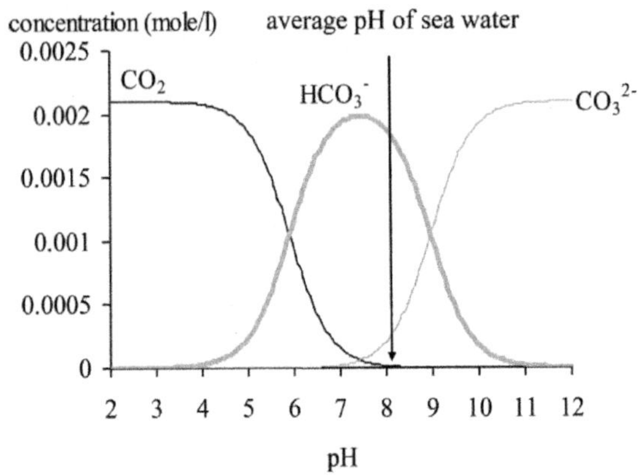

Figure 1.4. *Relative distribution of three chemical forms of dissolved inorganic carbon as a function of the pH in the average current conditions of seawater on the Earth's surface (total concentration of inorganic dissolved carbon or DIC = 2.1 mmole.kg^{-1}, salinity = 35 g of salts kg^{-1} of seawater, temperature = 25°C) (from [BER 08])*

COMMENTARY ON FIGURE 1.4.– Because of the general composition of dissolved compounds in seawater and with respect to electric neutrality, the current average pH of seawater is adjusted to around 8.1 to 8.2.

Evidently, periodic variations in the concentration of borate ion $B(OH)_4^-$ also play an important role in this adjustment and therefore have an effect on the fluxes of CO_2.

To summarize, the tectonic activity (either on land or on the ocean floor), with the ocean as chemical mediator, is the primary cause of long-term variations in the greenhouse effect. A lasting surplus in volcanic emissions of CO_2 with respect to the flux of CO_2 consumed by the alteration of extruded (mainly) basalt and other aluminosilicate rocks, favors the introduction of CO_2 into the ocean. This infiltration leads to an adjustment of the equilibriums of inorganic dissolved carbon with a slight increase in the acidity and a drop in alkalinity, which favors the CO_2 form of inorganic dissolved carbon and, through a physical exchange, increases the atmospheric level of CO_2, and therefore, the greenhouse effect. These modifications are only transient since the dissolving of sedimentary carbonates tends to cancel them out. For them to last, it is necessary that a relative excess of CO_2 emissions also lasts.

Conversely, a relative excess of weathering-related CO_2 consumption relative to volcanic emissions of CO_2 favors an increase in alkalinity due to the introduction of conservative cations dissolved in the ocean (see, for example, the formula [1.5] given above for the chemical weathering of anorthite), a drop in acidity and a reduction in the greenhouse effect. The modifications here are also fundamentally transient since they are compensated by the precipitation of solid carbonates. They are only long lasting if the disequilibrium between basalt extrusion and volcanic CO_2 emissions is maintained.

CO_2 emission resulting from the combustion of organic carbon (oil, gas, coal), in response to the energy demand of current human activity, is the main reason for the recent short-term increase in greenhouse effect. The increased rate of this process of destocking of organic fossil carbon evidently creates a great excess of CO_2 emissions relative to its consumption by the chemical weathering of aluminosilicate rocks. The long-term effects will be the same as those of the natural processes, although this imbalance starts with a short-term transitory accumulation of CO_2 in the atmosphere and has an immediate and very strong impact on the greenhouse effect.

Over the last few decades, after anthropogenic CO_2 started to be injected into the ocean, its direct effects, notably an increased ocean acidification, are clearly measurable [TUR 12]. More indirectly, this acidification has consequences for ecosystems where organisms that construct carbonate tests (small external shells) and other shells play an important role, such as coral reef systems.

On the scale of a few centuries after the stabilization or cessation of anthropogenic emissions of CO_2, the transitory disturbance of the greenhouse effect will cease. If emissions where to be stabilized, a new steady state of the Earth system will establish itself with a warmer global climate, slightly more acidic oceans, and communities biologically adapted to this new environment. If emissions where to cease, the perturbation will be compensated by the dissolving of sedimentary carbonates and the chemical weathering of aluminosilicate rocks with, perhaps, a return to the initial pre-industrial state. We note that this second hypothesis will occur sooner or later, when the reserves of fossil fuel are exhausted. Of course, this is theoretical and will only be true if the impacts of climatic variations that we are experiencing, or will experience, are strictly reversible. Indeed, many of them are probably not reversible, at least not in the short-term. This is the case with the release of methane during thawing permafrost, or with that of the deep methane hydrates. This is also the case with the positive retroaction linked to the decrease in the albedo of sea ice and seasonal snow coverings. Moreover, the combination of effects in a system as complex as the Earth system is almost unforeseeable (it is said that the system has a nonlinear behavior). It follows that a strict compensation for the anthropogenic disturbance of the greenhouse effect will perhaps not take place, and that the Earth system will attain another climactically steady state, different from the one that mankind knew before the industrial era.

1.5. Oceanic photosynthesis regulates itself on a short timescale

Section 1.4 has just described the regulation of the greenhouse effect, essentially through the regulation of the atmospheric level of CO_2 by the equilibriums of inorganic carbon dissolved in the ocean. The dissolved inorganic carbon, in the form of CO_2, is, moreover, the major nutrient for photosynthesis of which the simplified equation is shown in section 1.3. CO_2 itself is thus regulated through this process. This regulation has no direct impact on the biological productivity of the ocean since, as we will see, it is above all limited by nutritive

atmosphere. In the case of N_2 , the formula for its transformation is the following:

$$NO_3^- + 1{,}25CH_2O + H^+ \rightarrow 0{,}5N_2 + 1{,}25CO_2 + 1{,}75H_2O \qquad [1.9]$$

The compounds produced by denitrification, which contain nitrogen, are transported in dissolved form by oceanic circulation and molecular diffusion up to the surface of the ocean, where they are then in excess and cause a balancing flux into the atmosphere. The quantity of oceanic nitrate is thus diminished.

When the atomic ratio N/P in the interior ocean is greater than 16 (Redfield ratio), photosynthesis is not limited by the nitrate and, all things being equal, gives rise to a greater productivity and an increased exportation of organic matter into the interior ocean. The consequences are a development of anoxic environments, an increase in denitrification and, finally, a loss of nitrate from the ocean. This decreases the N/P ratio, thus readjusting the proportion of nitrate/phosphate to biological demand.

1.5.3. *The regulation of the N/P ratio*

Measurements made over a number of years in the diverse water masses of the current global ocean (e.g. [FOR 12]) show a statistical nitrate/phosphate ratio whose slope is close to 16, which is also the N/P ratio of the biological demand for phytoplanktonic productivity in the ocean.

For a long time, oceanographers wondered whether this similarity was due to the fact that the phytoplankton had adapted to the ocean's chemistry or whether, conversely, the chemistry resulted from the long-term action of the phytoplankton. Monitoring, over the course of the last 20 years, of the phenomena described above [ALT 95, AND 07, MAR 06] now enables us to confirm that both participants, the life forms and the biochemistry, permanently adjust to each other. This adjustment reacts especially to disturbances linked to climate change on timescales of the order of a few thousand years.

But this adjustment also reacts to more long-term evolutions. For example, the influx of phosphate and the ocean thermohaline circulation change, not only under the effect of climate changes, but also under tectonic and paleogeographic variations on timescales ranging from hundreds of millions of years to a few million years. Despite the regulation that we have just described, such modifications can lead the ocean to diverge slightly, but permanently, from the Redfield ratio N/P, thus creating conditions for a Darwinian evolution where new species are selected for, which promote a new Redfield ratio. The global data compiled in the NOAA Levitus Atlas in 1994 suggest an overall nitrate/phosphate ratio slightly under 16, which is perhaps evidence of such a transitory state. In the Earth's distant past, the variations in oxygen enrichment of the environment have themselves also contributed to variations in the Redfield ratio, either by reducing the extent of anoxic zones where the denitrification could occur or by modifying the average N/P ratio of the cell. For example, it is possible that the membrane arsenal, a major consumer of phosphorus, was developed and perfected in the course of evolution, leading to an increase in the demand for P by the organisms and, thus, to a decrease in the N/P ratio of the biological demand.

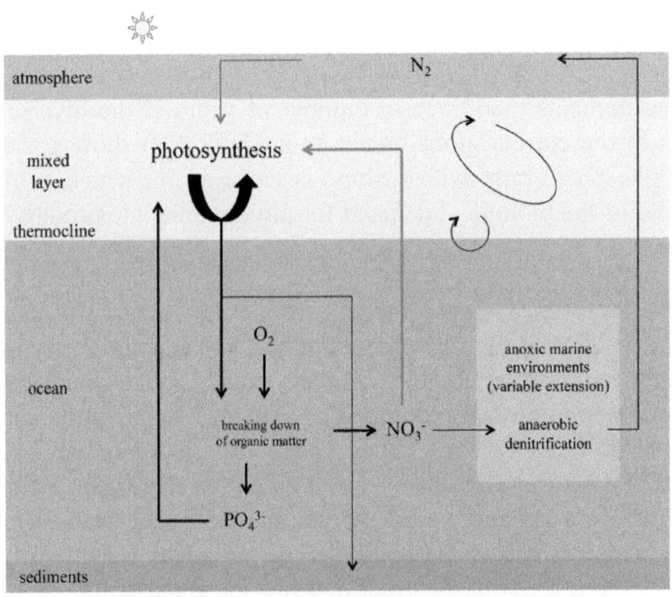

Figure 1.5. *The principle of the regulation of oceanic photosynthesis across the rapid cycle of major nutrients, nitrogen and phosphorus. (see color section)*

COMMENTARY ON FIGURE 1.5.– The majority of organic matter produced is recycled in the mixed layer through respiration in the trophic network. A small part is exported out into the interior ocean, where it is almost entirely degraded, always by respiration of the trophic network, whether it is deep pelagic or benthic. A very small fraction escapes degradation and fossilizes in the sediments, from where it is then integrated into the long-term geological cycle. When the nitrate is not limited relative to the phosphate, photosynthesis uses it preferentially and the flux of organic matter exported is at saturation, which consumes the oxygen in the inner ocean and creates anoxic environments. But these favor the denitrification and exportation of nitrogen into the atmosphere, thus reducing the oceanic stock of nitrate. If the nitrate is limited relative to the phosphate, photosynthesis calls upon species capable of fixing atmospheric nitrogen, whom then restock the oceanic nitrate through the process of exporting organic matter into the deep ocean and breaking it down.

1.6. Conclusion

1.6.1. *The ocean in the Earth system*

The few examples that have just been given, which are summarized in Figure 1.6, do not exhaust the complexity of the mechanisms of the Earth system. The previous section has shown the central role that the ocean plays in this system, notably through the following:

– plate tectonics and the related geological recycling of biogenic elements;

– the abiotic regulation of the greenhouse effect due to the chemical system of inorganic carbon dissolved in the ocean;

– the long-term increasing enrichment in oxygen O_2 of the environment and its self-regulation, both being due to the Darwinian evolution of respiratory metabolism in the ocean;

– the short-term self-regulation of the N/P ratio in the ocean due to oceanic circulation and the dynamic of phytoplankton species.

It would have been equally possible to examine other characteristics of the system that have played and still play a fundamental role in its evolution and regulations. Would life have existed without the magnetosphere protecting the Earth's surface from the Sun's ionizing radiation? What would the climate be like without the stability of the Earth's oblique axis of rotation, due to the existence of the Earth–Moon duo? How is the regulation of oceanic photosynthesis modulated by other nutritive elements, such as iron [PIC 09]?

But, of course, it requires more than a chapter, more than an entire book, to describe these phenomena in all their diversity and complexity. Interested readers will be able to find something to satisfy their curiosity in different individual or collected volumes [BER 07, BER 08, GAR 06, GAR 11, GAR 12, REI 06].

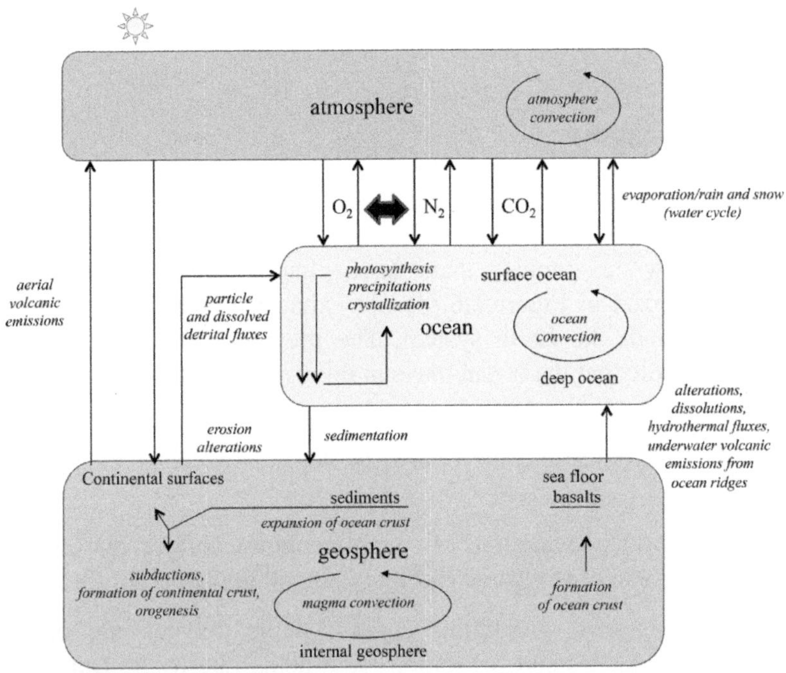

Figure 1.6. *The Earth system as it has functioned since the massive colonization of the ocean by photosynthetic life forms producing oxygen, that is to say for more than 2.5 billion years (see color section)*

COMMENTARY ON FIGURE 1.6.– Oxygen has only played an important biogeochemical role since its first global accumulation at the beginning of the Proterozoic Period (2.5 billion years ago). However, it is possible that it had a role previously (in the Archaean Period), by promoting the Darwinian selection of the first manifestations of aerobic respiratory metabolism in a few local environments where oxygen had already accumulated.

1.6.2. *The anthropogenic disturbance of the Earth system*

For two centuries, the demographic, technological and industrial development of human society has been such that it now has a global impact on the functioning of the Earth system. This is what is known as the 'anthropogenic disturbance'.

Certainly, this does not date from the industrial revolution since, from the Neolithic Period, man has modified his environment by deforestation and the mobilization of surfaces dedicated to agriculture and rearing livestock. Nevertheless, it is only from the end of the 19th Century that the global impact has started to be felt with the large-scale use of coal and then of petroleum and natural gas. This evolution very quickly accelerated in the course of the 20th Century with technological developments and the demographic explosion of the human population and its needs.

Technological progress is made available to a large proportion of people (electricity, automobiles, the personal computer, the mobile telephone, etc.), thus multiplying the industrial need for energy to make industrial products and meet individual energy needs for their use. In parallel, technological progress has enabled the mobilization of natural resources of solar fossil energy, through the combustion of coal and petroleum and natural gas, using oxygen from the atmosphere. The most recent technologies for exploration and exploitation now permit us to look for resources which, until recently, would have been impossible to access (e.g. deep oceanic environments, polar regions and shale gas), whereas the economic conditions of today render possible exploitation of reserves that were not, or were less, profitable a short time ago (e.g. bituminous shale and certain types of carbon). The direct consequences of this evolution are a massive destocking of organic fossil carbon and a degradation of the ecological status of the continental and marine environments.

Through combustion, we combine organic fossil matter with the oxygen from the atmosphere to form CO_2. This is in some way a gigantic biological respiration, even if it is not metabolic, with a result strictly inverse to that of photosynthesis. But this respiration uses the resource at a speed greatly exceeding that of its renewal. We consume in a few decades matter that accumulated in deposits (e.g. coal, petroleum and natural gas) over the course of tens of millions of years. We observe that while coal is a sedimentary rock initially rich in organic matter (mother rock), petroleum and natural gas accumulate in porous deposits (sandy or carbonated rocks, the so-called reservoir rocks) topped by impermeable layers of clay or salt deposits. These deposits can only be formed as the result of a long process in the course of which the mother rock reaches, by burying layer after layer over several kilometers, the conditions in temperature and pressure necessary for the formation of petroleum and natural gas (by natural thermal cracking), to their very slow expulsion from the mother rock (primary migration), and to their transfer and trapping in reservoir rocks (secondary migration). The use of shale gas implies a retrieval of gaseous hydrocarbons that have still not been expelled from the mother rock. It requires artificially creating porousness in the mother rock and a draining toward the surface to accelerate the expulsion.

The imbalance between the speed of exploitation and natural reconstitution of these resources does not affect the Earth system, but will oblige the human population itself to modify its way of life. It concerns the transition of energy, which is a major subject but which is outside the scope of this chapter. According to some thinkers and economists, such as Bourg and Jancovici [BOU 11, JAN 07], a lasting human civilization will oblige mankind to consume five times less energy than today.

The second impact is, of course, the increase in the concentration of greenhouse gases in the atmosphere, mainly CO_2 and CH_4

(methane), and the climatic consequences that follow: global warming, local modification of the climate, the melting of glaciers and ice caps, a rise in the average sea level, acidification of the oceans, possible destabilization of hydrates of methane in oceanic sediments at depth, positive retroaction by an increase in albedo, etc.

Despite its enormous power as a buffer, the ocean cannot regulate in the short-term the disturbance due to the current rate of anthropogenic greenhouse gas emissions. We are therefore in a transitional phase in the course of which the climate is modifying as a result of human activity. The consequences are already perceptible in numerous parts of the world. Beyond the immediate adaptation of human societies, man should also make sure not to modify the climate in an irreversible manner (threshold effect, irreversible melting of polar caps) so that a future way of energetically sustainable life can be established in habitable climatic zones as widespread as they are today.

On geological timescales, the Earth as a whole system, including the presence of life forms, will adapt itself and will pursue its existence, with or without mankind.

1.6.3. *And life among all that?*

In this chapter, the use of the term "life" has been carefully avoided, preferring that of "life forms" or "living things" or "living beings". As already underlined at the beginning (section 1.2.5), life is a concept much richer and more complex than the simple arbitrary definition that can be given to the term "alive" or "biological".

Before asking ourselves about the profound nature of life, it is sensible to ask what, among its characteristics, can differentiate it from "living".

As far as we presently know, terrestrial life has been uninterrupted for more than 3.5 billion years. This continuity is partly due to the phylogenetic link between all living beings. Life never ceases with the death of living beings; it is transmitted into new living beings. Even if a line of descent is interrupted, other branches continue to diversify in eventually using other ecological niches. Life is therefore an adaptive and colonizing phenomenon.

However, for life to be life, it requires that no living being is isolated. The isolation of a living being, whether that is for its reproduction, nutrition or respiration, means that it no longer participates in life during the time that separates it from its own death, which can be counted in thousands of years (certain trees) or in seconds.

Life therefore consists of all the evolutionary interactions and retroactions, whether between living beings themselves (trophic relations, cooperation, reproduction), or between living beings and the conditions in their environments (which are also evolving). Living beings are local and temporary self-regulating entities, which establish their own growth, cohesion, organization and the renewal of the matter from which they are constituted for the duration of their "life". However, we know that numerous sequences of DNA are transmitted totally or partially to their eventual descendants and this transmission is at the root of the ontogenesis of new living beings. Even if the deterministic impact of DNA on ontogenesis is strongly contested today by certain biologists [KUP 11], it constitutes no less than the initial information that the embryo of each living things receives. This information has been transmitted through the descendants of life forms for more than 3.5 billion years and it is this which evolves [DAW 11, GOU 02], even if its determinist role in the development of a given living creature (the "genetic code") is undoubtedly less important than it was thought during the years that followed the discovery of DNA. In the history of the Earth, the reproduction of a nucleic acid carrying genetic information (RNA having perhaps preceded DNA as the initial carrier of information) is therefore a characteristic of terrestrial life.

On a planet largely colonized by life forms, the plasticity of expression of phenotypes and the evolutionary plasticity of genotypes contribute together to adapt life forms to their non-biological environment. Reciprocally, this evolves in response to the pressure that these life forms exert on it. This ancient dynamic, more than 3.5 billion years old, is nothing other than life.

Our common life to which we contribute in this short instant!

1.7. Acknowledgments

The author very sincerely thanks the colleagues who have been happy to read and critique this work, first and foremost Louis Legendre, André Monaco and Patrick Prouzet.

1.8. Bibliography

[ALT 95] ALTABET M.A., FRANÇOIS R., MURRAY D.W., *et al.*, "Climate-related variations in denitrification in the Arabian Sea from sediment $^{15}N/^{14}N$ ratios", *Nature*, vol. 373, pp. 506–509, 1995.

[AND 07] ANDERSON T.R., RYABCHENKO V.A., FASHAM M.J.R., *et al.*, "Denitrification in the Arabian Sea: a 3D ecosystem modeling study", *Deep Sea Research Part I: Oceanographic Research Papers*, vol. 54, no. 12, pp. 2082–2119, 2007.

[BER 04] BERNER R.A., *The Phanerozoic Carbon Cycle, CO_2 and O_2*, Oxford University Press, New York, 2004.

[BER 07] BERTRAND P., "Towards a global Earth's regulation", in GARGAUD M., MARTIN H., CLAEYS P. (eds.), *Lectures in Astrobiology II*, Springer-Verlag, Berlin Heidelberg, pp. 281–302, 2007.

[BER 08] BERTRAND P., *Les attracteurs de Gaïa*, Editions Publibook Université, Paris, 2008.

[BIB 09] BIBRING J.P., *Mars, planète bleue?*, Odile Jacob, Paris, 2009.

[BIL 01] BILLEN M.I., GURNIS M., "A low viscosity wedge in subduction zones", *Earth and Planetary Science Letters*, vol. 193, nos. 1–2, pp. 227–236, 2001.

[BOU 11] BOURG A., *Pour une 6e république écologique*, Odile Jacob, Paris, 2011.

[CAN 07] CANFIELD D.E., POULTON S.W., NARBONNE G.M., "Late-neoproterozoic deep-ocean oxygenation and the rise of animal life", *Science*, vol. 315, pp. 92–95, 2007.

[CAT 05] CATLING D.C., CLAIRE M.W., "How Earth's atmosphere evolved to an oxic state: a status report", *Earth and Planetary Science Letters*, vol. 237, pp. 1–20, 2005.

[DAW 11] DAWKINS R., *The Blind Watchmaker*, PP Science, Penguin Books, London, 2011.

[DIC 81] DICKSON A.G., "An exact definition of total alkalinity and a procedure for the estimation of alkalinity and total inorganic carbon from titration data", *Deep-Sea Research*, vol. 28A, pp. 609–623, 1981.

[DIX 04] DIXON J.E., DIXON T.H., BELL D.R., *et al.*, "Lateral variation in upper mantle viscosity: role of water", *Earth and Planetary Science Letters*, vol. 222, no. 2, pp. 451–467, 2004.

[DON 04] DONNADIEU Y., GODDÉRIS Y., RAMSTEIN G., *et al.*, "A 'snowball earth' climate triggered by continental break-up through changes in runoff", *Nature*, vol. 428, pp. 303–306, 2004.

[ELA 10] EL ALBANI A., BENGTSON S., CANFIELD D.E., *et al.*, "Large colonial organisms with coordinated growth in oxygenated environments 2.1 Gyr ago", *Nature*, vol. 466, pp. 100–104, 2010.

[ESA 09] ESA, "New map hints at Venus' wet, volcanic past", 2009. Available at www.esa.int/esaMI/Venus_Express/SEMUQCLXOWF_0.html.

[FOR 12] FORRYAN A., ALLEN J.T., EDHOUSE E., *et al.*, "Turbulent mixing in the eddy transport of Western Mediterranean intermediate water to the Alboran Sea", *Journal of Geophysical Research*, vol. 117, p. C09008, 2012.

[FRO 82] FROELICH P.N., BENDER M.L., LUEDTKE N.A., *et al.*, "The marine phosphorus cycle", *American Journal of Science*, vol. 282, pp. 474–511, 1982.

[GAR 06] GARGAUD M., CLAEYS P., LOPEZ-GARCIA P., *et al.*, *From Suns to Life: A Chronological Approach to the History of Life on Earth*, Springer, Dordrecht, reprinted from *Earth Moon, and Planets*, vol. 98, nos. 1–4, 2006.

[GAR 11] GARGAUD M., AMILS R., CERNICHARO J., *et al.*, *Encyclopedia of Astrobiology*, Springer, Berlin, 2011.

[GAR 12] GARGAUD M., LOPEZ-GARCIA P., MARTIN H., *et al.*, *Young Sun, Early Earth and the Origins of Life*, Springer-Verlag, Berlin, Heidelberg, 2012.

[GOU 02] GOUYON P.H., HENRY J.P., ARNOULD J., *Gene Avatars, the Neo-Darwinian Theory of Evolution*, Kluwer Academic/Plenum Publishers, New York, 2002.

[HOF 98] HOFFMANN P.F., KAUFMAN A.J., HALVERSON G.P., *et al.*, "A neoproterozoic snowball earth", *Science*, vol. 281, pp. 1342–1346, 1998.

[HOF 05] HOFMANN A., "The geochemistry of sedimentary rocks from the Fig Tree Group, Barberton greenstone belt: implications for tectonic, hydrothermal and surface processes during mid-Archaean times", *Precambrian Research*, vol. 143, pp. 23–49, 2005.

[HOL 84] HOLLAND H.D., *The Chemical Evolution of the Atmosphere and Oceans*, Princeton University Press, Princeton, 1984.

[JAN 07] JANCOVICI J.M., GRANJEAN A., *Le plein s'il vous plaît*, Le Seuil, Paris, 2007.

[KEI 04] KEIM C.N., MARTINS J.L., ABREU F., *et al.*, "Multicellular lifecycle of magnetotactic prokaryotes", *FEMS Microbiology Letters*, vol. 240, no. 2, pp. 203–208, 2004.

[KOK 00] KOKUBO E., IDA S., MAKINO J., "Evolution of a circumterrestrial disk and formation of a single moon", *Icarus*, vol. 148, pp. 419–436, 2000.

[KUP 12] KUPIEC J.J., "Constrained stochasticity and cellular Darwinism (stochastic gene expression in cell differentiation and embryo development)", *SciTopics*, 11 September 2012. Available at www.scitopics.com/Constrained_stochasticity_and_cellular_Darwinism_stochastic_gene_expression_in_cell_differentiation_and_embryo_development.html.

[LIU 04] LIU L.G., "The inception of the oceans and CO_2-atmosphere in the early history of the Earth", *Earth and Planetary Science Letters*, vol. 227, pp. 179–184, 2004.

[MAR 97] MARGULIS L., SAGAN D., *Microcosmos: Four Billion Years of Microbial Evolution*, University of California Press, 1997.

[MAR 06] MARTINEZ P., LAMY F., ROBINSON R.R., *et al.*, "A typical $\delta^{15}N$ variations at the southern boundary of the East Pacific oxygen minimum zone over the last 50 ka", *Quaternary Science Reviews*, vol. 25, pp. 3017–3028, 2006.

[MAS 03] MASSABUAU J.C., "Primitive, and protective, our cellular oxygenation status?", *Mechanisms of Ageing and Development*, vol. 124, nos. 8–9, pp. 857–863, 2003.

[MON 14b] MONACO A., PROUZER P., (eds.), *The Land-Sea Interactions*, ISTE, London and John Wiley & Sons, New York, 2014.

[PIC 09] PICHEVIN L.E., REYNOLDS B.C., GANESHRAM R.S., *et al.*, "Enhanced carbon pump inferred from relaxation of nutrient limitation in the glacial ocean", *Nature*, vol. 459, pp. 1114–1117, 2009.

[POP 62] POPPER K.R., *Conjectures and Refutations: The Growth of Scientific Knowledge*, Basic Books, New York, 1962.

[RAS 08] RASMUSSEN B., FLETCHER I.R., BROCKS J.J., *et al.*, "Reassessing the first appearance of eukaryotes and cyanobacteria", *Nature*, vol. 455, pp. 1101–1104, 2008.

[RED 58] REDFIELD A.C., "The biological control of chemical factors in the environment", *American Science*, vol. 46, pp. 205–221, 1958.

[REI 06] REISSE J., *La longue histoire de la matière, une complexité croissante depuis des milliards d'années*, PUF, Paris, 2006.

[REY 01] REYSENBACH A.L., CADY S.L., "Microbiology of ancient and modern hydrothermal systems", *Trends in Microbiology*, vol. 9, no. 2, pp. 79–86, 2001.

[RUI 11] RUIZ J., "Giant impacts and the initiation of plate tectonics on terrestrial planets", *Planetary and Space Science*, vol. 59, no. 8, pp. 749–753, 2011.

[RUI 07] RUIZ-TRILLO I., BURGER G., HOLLAND P.W.H., *et al.*, "The origins of multicellularity: a multi-taxon genome initiative", *Trends in Genetics*, vol. 23, no. 3, pp. 113–118, 2007.

[SAC 08] SACHS J.L., "Resolving the first steps to multicellularity", *Trends in Ecology & Evolution*, vol. 23, no. 5, pp. 245–248, 2008.

[SCH 10] SCHAEFER L., FEGLEY JR. B., "Chemistry of atmospheres formed during accretion of the Earth and other terrestrial planets", *Icarus*, vol. 208, pp. 438–448, 2010.

[SCH 91] SCHLESINGER W.H., *Biogeochemistry: An Analysis of Global Change*, Academic Press, San Diego, 1991.

[SEI 93] SEITZ A.P., NIELSEN T.H., OVERMANN J., "Physiology of purple sulfur bacteria forming macroscopic aggregates in Great Sippewisset Salt Marsh, Massachusetts", *FEMS Microbiology Ecology*, vol. 12, no. 4, pp. 225–235, 1993.

[SHO 96] SHOCK E.L., "Hydrothermal systems as environments for the emergence of life", *Ciba Foundation Symposium 202 – Evolution of Hydrothermal Ecosystems on Earth (and Mars?)*, John Wiley & Sons, Chichester, 1996.

[TUR 12] TURLEY C., GATTUSO J.P., "Future biological and ecosystem impacts of ocean acidification and their socioeconomic-policy implications", *Current Opinion in Environmental Sustainability*, vol. 4, pp. 278–286, 2012.

[TYR 99] TYRELL T., "The relative influence of nitrogen and phosphorus on oceanic primary production", *Nature*, vol. 400, pp. 525–531, 1999.

[VOE 10] VOEGELIN A.R., NÄGLER T.F., BEUKES N.J., *et al.*, "Molybdenum isotopes in late Archean carbonate rocks: implications for early Earth oxygenation", *Precambrian Research*, vol. 182, pp. 70–82, 2010.

[WHI 80] WHITTINGTON H.B., "The significance of the fauna of the Burgess Shale, Middle Cambrian", *British Columbia, Proceedings of the Geologists' Association*, vol. 91, no. 3, pp. 127–148, 1980.

[WIL 01] WILDE S.A., VALLEY J.W., *et al.*, "Evidence from detrital zircons for the existence of continental crust and oceans on the Earth 4.4 Gyr ago", *Nature*, vol. 409, pp. 175–178, 2001.

[ZEE 01] ZEEBE R.E., WOLF GLADROW D., "CO_2 in seawater: equilibrium, kinetics, isotopes", *Elsevier Oceanography Series*, vol. 65, 2001.

2

The Ocean and the Climate System

2.1. Introduction

Global warming, now well attested on a global scale, remains an important subject of study since its characteristics are difficult to typify, on account of its complexity and the inadequacy of our current knowledge of the state of the ocean. Anticipating how the intrinsic properties that characterize the ocean contribute to shaping the state of the climate and its evolution remains a major challenge. After presenting some findings concerning the changes already observed and methods for studying future projections, we will return to the behavior of the ocean and the way in which it could influence the evolution of the climate by its particular dynamics and rhythm.

2.2. Climate change

The Earth has experienced greatly contrasting climates, from its "snowball" stage, 640 million years ago when the average temperature was about –50°C, to the very hot climate of the Cretaceous Period, when the average temperature exceeded 20°C. More recently, the last few million years have been marked by an alternation of ice ages interspersed with hot interglacial periods. These variations have been attributed to changes in the astronomic parameters that influence the distribution of solar energy received by the Earth's surface. For the

---

Chapter written by Pascale DELECLUSE.

last eight thousand years, the Earth has experienced a remarkably stable and hot period, with an average temperature of 15°C that should persist for quite a long time given the stable configuration of its current orbit's astronomical parameters. However, this equilibrium is brought into question by the increase in pressures resulting from human activity. These pressures are visible in numerous studies that agree on a global warming of the temperature of the Earth's surface. In view of this situation, an intergovernmental group, the IPCC[1], has been created through the initiative of the WMO[2] and UNEP[3], in order to regularly carry out an assessment exercise on the climatic conditions of the Earth's surface. The majority of the results discussed below are taken from reports from IPCC and the literature that has contributed to them.

2.2.1. *The report on the findings*

The combination of long series of temperature measurements on the ground, sea and more recently from space, allows us to plot a global distribution of the changes in surface temperature. Several centers contribute to this collective effort to validate and analyze the data, including the *Climate Research Unit* (CRU) with the Hadley Center in the Met Office (HadCRUT3), the *National Climate Data Center* (NCDC) and the *Goddard Institute for Space Studies* (GISS). These long series are homogenized to correct instrument errors, changes in the observation sites, disturbances linked to the extension of urban heat islands, etc. They are also spatialized, using spatial observations to compensate for lacunae in *in situ* coverage. The picture that we draw from these time series is very coherent. The conclusion of the IPCC AR4[4] [SOL 07] is that the Earth's temperature has increased by 0.76 [±0.19]°C since the pre-industrial era (1850–1899) and that the speed of this warming has progressively accelerated, the degree of warming having doubled in the last fifty years in comparison with the last century. The years 2001–2007 have been particularly hot, with an

1 IPCC: Intergovernmental Panel on Climate Change.
2 WMO: World Meteorological Organization.
3 UNEP: United Nations Environment Programme.
4 AR4: The IPCC fourth assessment report.

average anomaly of 0.54°C relative to an average calculated over the period 1951–1980. The year 2008 was certainly a little cooler than the years that preceded it, but the increase was still 0.44°C (see http://data.giss.nasa.gov/gistemp/). Since then, global temperatures seem to be stagnating, but it is still too early to determine the exact reasons for this behavior: does this slowing represent a manifestation of the climate's natural variability, is this a strengthening of the trapping of heat by the ocean, or of the charge in aerosols? The causes of this slowing could be multiple after the marked acceleration in the last decade of the 20th Century and the beginning of the 21st Century. We can see in Figure 2.1 that the warming has not followed a constant trajectory and that its progression has been marked by numerous plateaus.

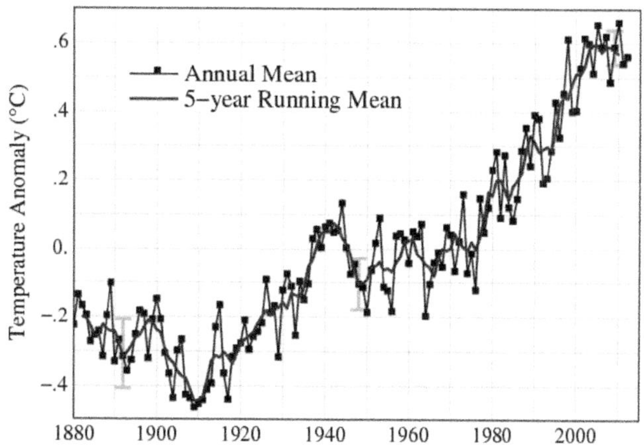

Figure 2.1. *Surface temperature as a global average for land and sea since 1880 (data from http://data.giss.nasa.gov/gistemp/)*

The regional distribution is characterized by warm patterns on the continents at high latitudes, the warming being less at low latitudes and over the ocean. Figure 2.2 shows the distribution of anomalies in temperature over two periods of five years, 1998–2002 and 2008–2012 compared to an average distribution over (1951–1980), calculated from data and software available on the site http://data.giss.nasa.gov/gistemp/. Signs of intensified warming at high latitudes of the northern hemisphere are very noticeable. We note that even the southern hemisphere shows moderate but significant patterns. It is however necessary to calculate averages over a sufficiently long period of time for a mean figure to emerge since the large natural variability from one year to another can mask signs of warming.

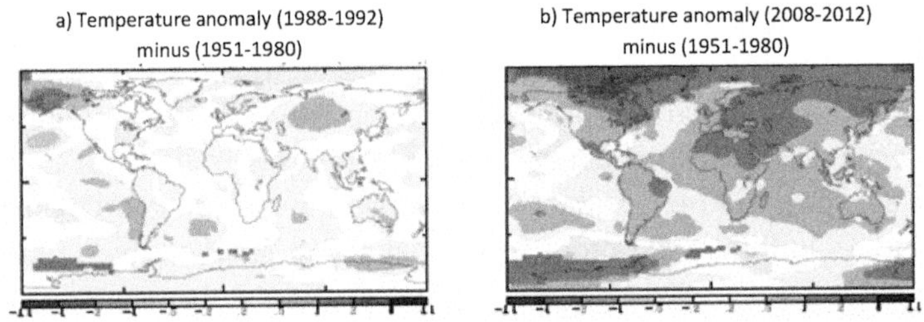

Figure 2.2. *Distribution of anomalies in temperature on the Earth's surface for 1988–1992 and 2008–2012 compared to the reference period 1951–1980 (according to http://data.giss.nasa.gov/gistemp/) (see color section)*

This warming of the land's surface manifests itself in numerous effects such as the decrease in snow coverage in winter as well as the almost universal melting of glaciers, at low latitudes as well as at temperate latitudes. The rapid melting of the sea ice is equally manifest and its extent has constantly diminished to reach a minimum coverage in September 2007, when it was reduced to 39% less than the average over 1979–2000, a record reached again in 2012.

The world's oceans are warming and the warm patterns, initially observed in the surface layers, are slowly penetrating the interior where they are observed up to a depth of 3,000 m in some regions. However, the distribution of the warm patterns is far from being homogeneous since it depends on the overall circulation of the oceans, which we will discuss later.

2.2.2. *Interpretation of the observed changes*

The Earth's climate is far from being stable and its history shows a complex evolution. Multiple factors shape it and contribute to its evolution on different time scales. The shape of continents, for example, conditions the circulation of the atmosphere and ocean. The chemical composition of the atmosphere also reflects the slow evolution of the lithosphere and biosphere and contributes to characterizing the climate. On a shorter scale of time, dust emissions due to violent volcanic eruptions that can reach the stratosphere lead to a cooling of the Earth's surface for a few months to a few years. On the scale of the last million years, extensive information taken from polar ice cores and from oceanic sediments demonstrates the importance of the parameters controlling the Earth's orbit around the Sun. If the origin of these oscillations between a dominant glacial climate state and hot interglacial periods is easily identifiable in the difference in distribution of the solar energy received by the Earth's surface, the amplitude of these oscillations can only be explained by implying feedback between greenhouse gases and the climate, variations in solar energy being much too weak to explain the size of the observed variations.

These strong variations in the land climate have been accompanied by changes in polar caps and in sea level. The analysis of air bubbles trapped in polar cores enable the reconstruction of atmospheric concentrations of trace gases; they show important variations with an incidence of elevated concentration rates of carbon dioxide (CO_2), methane (CH_4) and nitrogen protoxide (N_2O) during interglacial periods and low levels in glacial periods. Moreover, these records indicate that current levels of CH_4 (1,774 ppb[5] in 2006) and CO_2 (379 ppm[6] in 2006) had never been reached during the last million years, which suggests that the current climate has not experienced a parallel during this period.

2.2.3. *The Earth's radiative equilibrium and greenhouse gases*

The Earth is in a delicate equilibrium between the energy received from the Sun and the energy that it emits into space. It is the small variations in this distribution of energy on the Earth's surface that paleoclimate specialists held responsible for oscillations between

5 ppb: part per billion in volume.
6 ppm: part per million in volume.

glacial and interglacial climates. The Sun, which has a high temperature in the order of 6,000 K emits radiation in the band X and visible spectrum whereas the Earth, whose average temperature is 255 K, mainly emits radiation in infrared wavelengths following Stefan–Boltzmann's law of blackbody radiation [TRE 92]. To say that the Earth is in a radiative equilibrium means that the energy received from the Sun is equal to the energy emitted into space, which leads to an average "radiative" temperature of 255 K (or –18°C). Indeed, the average temperature observed is close to 15°C, or 33°C more than the radiative temperature. This gap is due to the presence of gas in very small quantities (trace gases) whose interaction with different radiative wavelengths significantly alters the energy balance, in quantity according to the wavelength. Of these gases, greenhouse gases in particular react with infrared wavelengths.

The main greenhouse gas is water vapor. Water, whether it be in vapor, solid or liquid form, is indeed the main contributor to the warming of atmospheric conditions (more than 60%). Water vapor evidently contributes to the character and evolution of the atmosphere; its distribution is very heterogeneous, its renewal is rapid and the course of its cycle plays an active role in the equations that describe the atmosphere. Other significant contributions to the greenhouse effect are brought about by trace gases with a long life (CO_2, CH_4, N_2O, etc.) just as by more reactive components such as ozone. These contributions vary according to the properties of radiative absorption of the gas considered and according to their concentration.

2.2.4. *The role of greenhouse gases – GHG*[7]

The evolution of human activity, through the exploitation of soils, agriculture, industry and transport, is built upon an intensive exploitation of natural resources; the production of energy, initially drawn from coal, has progressively moved toward the exploitation of other forms of fossil hydrocarbons. The consumption of fossil energy, deforestation and intensive agriculture have produced a lot of trace

7 GHG: greenhouse gas.

gases, some are very volatile but others have a long life and their long residence time in the atmosphere enables homogenization on a global scale. The residence time of CH_4, for example, is in the order of 12 years, N_2O more than a hundred years and CO_2, chemically inert, only disappears from the atmosphere through absorption in the land biosphere and in the ocean. Well mixed in the atmosphere, these gases affect its radiative capacity in a durable way. Their signature is therefore visible on a global scale, even in isolated sites such as the Austral Islands in the Indian Ocean or in Antarctica. "Radiative forcing" measures how the energy present in the global system is altered by the presence of a gas which interacts with solar radiation and the radiation emitted by the Earth and atmosphere: it is defined as the net change in radiation at the summit of the lower atmosphere after having allowed for the upper atmosphere to readjust its radiative equilibrium, the temperatures of the surface and the troposphere remaining fixed at their nondisturbed value. This idea of radiative forcing allows us to evaluate and compare potential players in climate change in the same unit: CO_2 emerges as the most important player with a radiative forcing of 1.66 Wm^{-2}, CH_4, which has a stronger radiative forcing potential but whose concentration and life span are weaker is second at 0.48 Wm^{-2}, then follow the other gases, like N_2O at 0.16 Wm^{-2}.

The concentration of CO_2 has increased by 36% from 280 ppm in the pre-industrial era to 379 ppm in 2006, and in 2013 approached 400 ppm at the site of Mauna Loa in Hawaii (www.esrl.noaa.gov/gmd/ccgg/trends/), a level of concentration that had never been observed in the previous 600,000 years recorded in the polar cores of Dome C in Antarctica. The increase in the level of concentration however only represents half of the level of emissions: almost half the emissions resulting from human activity are indeed absorbed by the natural cycle of CO_2 in the land and marine biosphere through photosynthesis and by their solubility. The physiology of plants allows the fixing of some of the CO_2 excess in the environment. However, this stimulation is not sustainable in the long term for many ecosystems and it is to be feared that some of them cannot adapt to new environmental conditions, which would lead to the disappearance of some biotopes, thus reinforcing the local greenhouse effect in some regions (for example this is what occurred in France during the heat wave of 2003).

The question of what climatic warming is attributable to is crucial and has been explored with the help of coordinated experiments with numerical simulation. To do this, numerical climate models are used, which describe the evolution of the two fluids, the ocean and the atmosphere, taking account of their exchanges with ice and vegetation, under the imposed boundary conditions. An initial group of experiments is carried out by assessing the evolution conditions of solar forcing and volcanic eruptions since 1850. Then a second group of experiments follows the same experimental protocol with the addition of the radiative contribution of GHG and aerosols. We note here the important work of coordinated experiments that has been in place for the last 15 years within the framework of the *Climate Modeling Intercomparison Program* (CMIP) group of the World Climate Research Programme. The climate models used over the last 20 years have steadily improved over the course of time. Although they show the main characteristics of regional climates, biases with observation persist in certain regions and show the difficulty of correctly representing certain phenomena. These biases are difficult to reduce, notably because they often result from interactions between different processes. For example, the choice of parameters to represent clouds can vary from one modeling group to another, leading to significant regional differences. Moreover, the atmosphere and the ocean are turbulent fluids and divergent trajectories can result rapidly from the behavior of small structures. To work around these difficulties, we do not only consider one simulation but rather an ensemble of several simulations carried out with slight alterations in their initial state or in the choice of parameters. Each model produces a set of several simulations of which it is hoped the average follows the observed trajectory. By combining the average trajectories of several models, it is possible to estimate the range of uncertainty resulting from the biases of the models. This work relies on the coordination of modeling teams at international level.

Comparing the average temperature trajectory in these ensembles both globally and regionally, the analyses within the framework of the IPCC AR4 report have shown that for the last 30 years, the Earth has been warming to a degree that is more marked in the northern hemisphere than in the southern hemisphere and that the spatial and temporal imprint of this warming can only be obtained when the additional radiative forcings, due to human activity, are imposed.

2.2.5. *Scenarios and projections*

Once the conclusion that the current warming is mainly due to human activity has been drawn, the question which arises is how we ascertain the climate toward which the Earth will evolve during the current century. To do this, it is necessary to make some hypotheses on the future emissions of GHG and therefore on the evolution of human societies. Integrated models (demography – economy – evolution) have been created to try to predict the emissions over the century to come while taking account of hypotheses for contained or drastically increasing demographic evolution, for economic evolution, for globalization or on the contrary a reduction in global interaction, for the emergence of new technologies, for an environmental conscience (either strong or nonexistent), etc. The combination of these hypotheses enables the construction of scenarios for the consumption of GHG and therefore of emissions for the 21st Century (see Figure 2.3) which have been called the *Special Report on Emission Scenarios* scenarios (for the IPCC SRES).

For a given SRES scenario, the predicted emissions are converted into concentrations of GHG, then these concentrations are imposed as external forcing on an ensemble of coupled ocean–atmosphere models of different complexity, the protocol for intercomparison being decided in the framework of the CMIP. It is in this way that a set of projections reaching to the end of the 21st Century and sometimes beyond has been created. A limited number of SRES scenarios have been selected to explore some contrasting trajectories:

– scenario SRESB1, the most optimistic, predicts a rapid evolution toward a service and information economy, leading to global solutions to ensure the eco-responsible evolution of societies;

– scenario SRESA2 describes a heterogeneous world, privileging regional economic growth;

– scenario SRESA1F1 describes a world of rapid economic growth relying on technological evolutions that permit the intensive exploitation of fossil fuels.

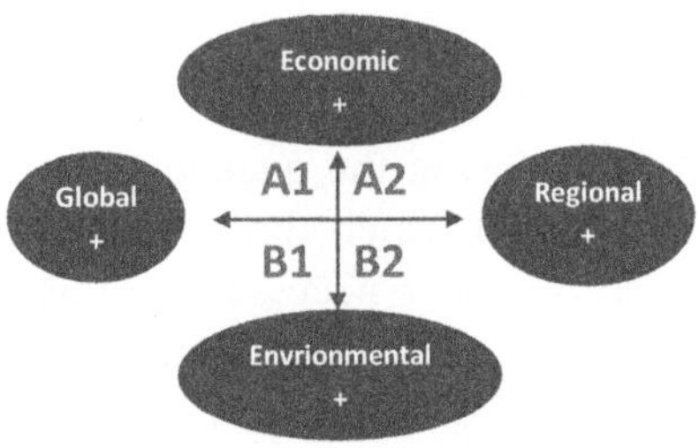

Figure 2.3. *Schematic representation of weights of hypotheses used in SRES scenarios*

This strategy has been deeply modified following the IPCC AR4 report, to permit research groups working on climate physics and those working on socio-economic modeling to work in parallel. The idea is now to work with the imposed trajectories of concentration, called RCP for the *Representative Concentration Pathway*, which have been proposed by Moss *et al.* [MOS 08] and which now serve as conditions of external forcing for climate models as well as for socio-economists who need to construct the conditions of societal evolution which will allow these trajectories to be reached.

The construction of RCP scenarios is largely inspired by the SRES scenarios used previously. Their name is determined by the value of total radiative forcing from GHG emissions reached in 2100 (see Figure 2.4). Thus, trajectory RCP8.5 corresponds to the strongest radiative forcing. It reaches 8.5 W/m^2 in 2100 and then increases. It corresponds to scenario SRESA2 and is considered the upper limit of the range for emissions; scenario RCP4.5, which approximately follows the trajectory of SRESB1, reaches a radiative forcing of 4.5 W/m^2 in 2100 and then stabilizes around this level. Scenario RCP2.6, after an excess predicted around 2030–2040, seeks to impose measures for mitigation that will enable the reduction of emissions to meet the policy objective not to exceed 2°C of global warming. The choice of these trajectories, expressed in radiative forcing, is directly usable by coupled models in their external forcing; the coherence of the RCP trajectories with the SRES enables the comparison of new results with old simulations; the RCP trajectories permit us to take account of policies for mitigation, without being prescriptive, and of socio-economic models for picturing measures to take in order to follow a given trajectory.

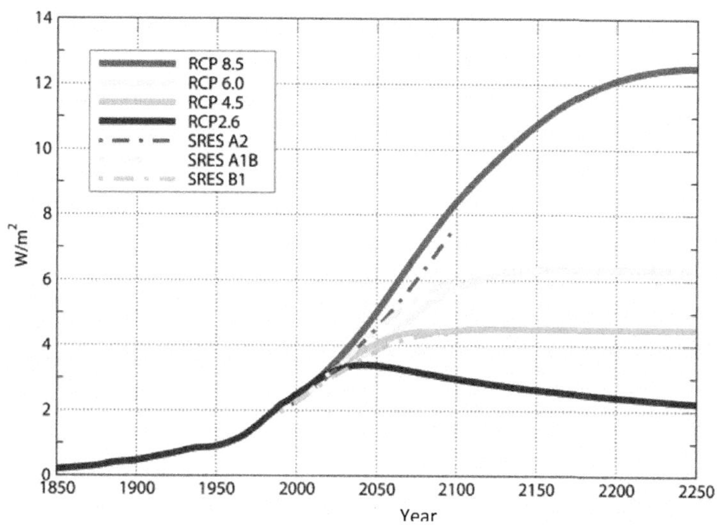

Figure 2.4. *Evolution of the Earth's radiative budget or "radiative forcing", in Wm^{-2} for the period 1850–2250. From 2006, the continuous lines represent new scenarios called Representative Concentration Pathways; the dotted lines represent the old scenarios (IPCC, 2007 and 2001) (© Stéphane Senesi (CNRM-GAME, Météo-France/CNRS)) (see color section)*

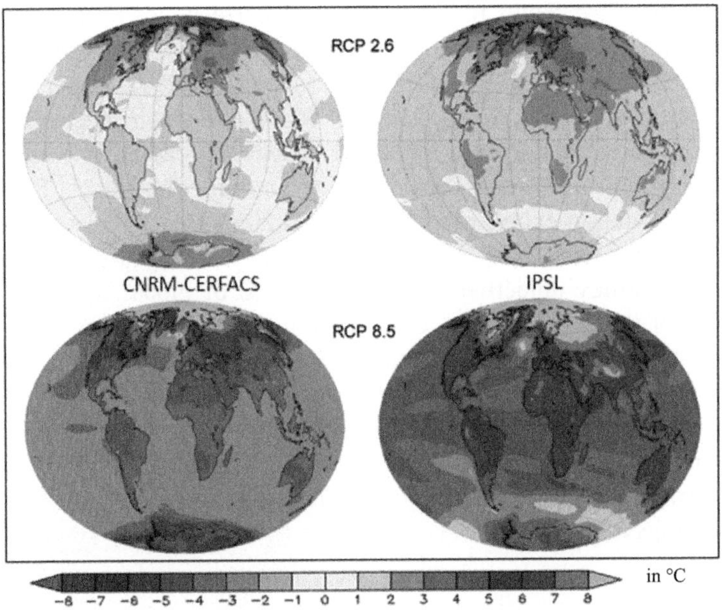

Figure 2.5. *Projection of the global temperature anomalies for 2071–2100 compared to 1961–1990 for model CNRM-CERFACS on the left and model IPSL on the right, for RCP 2.6 above and RCP 8.5 below (©Patrick Brockmann (LSCE/IPSL, CEA/CNRS/UVSQ)) (see color section)*

For the group of coupled models present in IPCC AR4, the temperature rises by 1.7 [1.0–2.7]°C for scenario SRESB1 and by 4.0 [2.4–6.3]°C for scenario SRESA1F1 at the end of the 21st Century. The trajectories are almost indiscernible until 2040 but nevertheless, they deviate greatly depending on the scenario for the second half of the century, underlining the importance of long-term necessary measures for mitigation. In spite of the fairly large range between the projections, the spatial behavior of the warming is very coherent for the overall set of models: the warming intensifies depending on latitude; it is weaker in equatorial regions and increases towards the poles; moreover, it intensifies above large continental masses. Figure 2.5 shows the distribution of anomalies in temperature on the Earth's surface, averaged between 2071 and 2100 minus the average temperature for 1961 to 1990, for the two French models that are included in the set of CMIP models used in the IPCC report, and for two trajectories RCP2.6 and RCP8.5. The size of the anomaly is clearly dependent on the scenario from 1 to 3°C with RCP2.6 on France, to 3–5°C with RCP8.5 for the model CNRM-CERFACS. We note, however, that at high latitudes, the maximum warming is situated in the Arctic for CNRM-CERFACS whereas model IPSL-CM extends it over Asia. This regional difference illustrates the critical impact that the regions covered by ice represent, regions where strong feedbacks between snow/ice and the climate occur, and where slight variations in the choice of parameters can lead to divergences.

In terms of precipitation, which is an important parameter for characterizing a regional climate, it is more difficult to obtain a consensus from an analysis of the models since the structure of this field is very heterogeneous spatially and sensitive to slight differences between models. Nevertheless, some common patterns appear in the structure of the anomalies. The anomalies in precipitation in the French models CNRM-CERFACS and IPSL-CM are presented in Figure 2.6. We can note in general an increase in precipitation in regions already notable for heavy precipitation (like the intertropical convergence zone in the Pacific or the north of Europe). However, decreases in precipitation are making themselves felt in regions that are already dry, such as the Mediterranean basin. The evolution of precipitation reflects the intensification of the hydrological cycle, associated with an increase in the atmosphere's capacity to retain water; the average residence time of water vapor in the atmosphere has increased.

As we see in this succinct description of the observations on the changes underway and the projections from coupled simulation experiments, the climate system's response is complex and involves numerous phenomena, on different time scales. We will now concentrate our attention on the ocean in order to clarify its role in the climate system.

2.3. Physics and dynamics

Oceans cover 70% of the Earth's surface and hold 97% of its water. Of the two intricately interlinked fluids, oceans and atmosphere, which evolve to balance the Earth's climate, the oceans play the role of a slow partner, whereas the atmosphere is the quick partner. In temperature for example, the oceanic extremes go from –1.6°C in polar regions, where sea ice forms (as continental temperatures are less than –50°C), to 32°C in the semi-enclosed tropical seas, such as the Red Sea, far removed from the extremes of the Sahara. The oceans act as a moderator for extremes of atmospheric temperature.

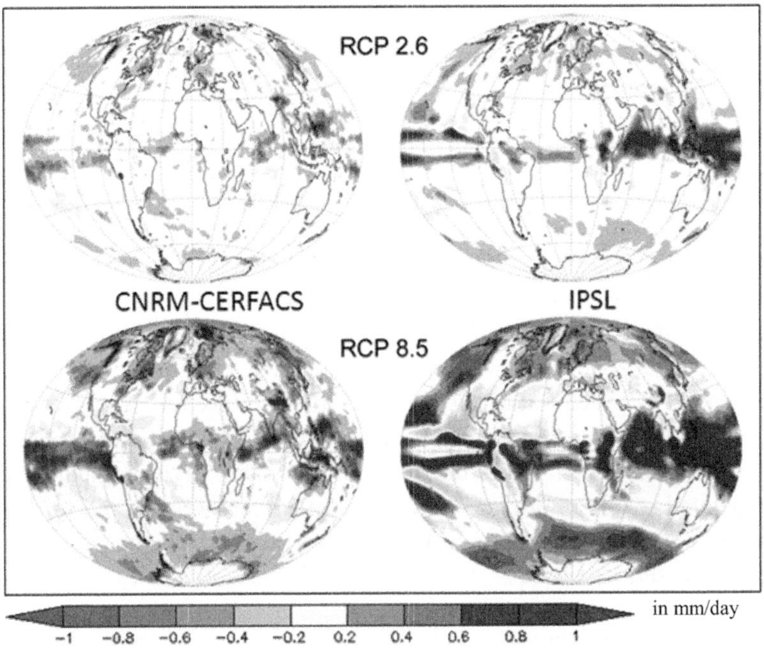

Figure 2.6. *Changes in precipitation in the period 2071–2100 compared to those in the period 1971–2000 calculated from models by CNRM-CERFACS and IPSL, for RCP2.6 above and RCP8.5 below (©Patrick Brockmann (LSCE/IPSL, CEA/CNRS/UVSQ)) (see color section)*

The other variable that characterizes seawater is its salt content. Its quantity is measured in grams of salt per kilogram of water. The saltiest waters are concentrated in the semi-enclosed tropical seas, such as the Red Sea where intense evaporation helps to concentrate the salt (40 g of salt/kg) whereas in the semi-enclosed seas close to the

Arctic, such as the Bering Sea, the waters are very fresh (20 g of salt/kg).

One characteristic property of a fluid is its volumic mass. That of the ocean depends on three quantities: temperature, salinity and pressure. The volumic mass of seawater varies between 1,022 and 1,029 kg/m^3 with an average value of 1,025 kg/m^3, which is around 1,000 times the volumic mass of air. The anomaly of volumic mass is usually introduced, which corresponds to the difference in volumic mass between seawater and fresh water: $\sigma = \rho - \rho_0 = \rho - 1{,}000$ kg/m^3. This quantity is generally called "density", surfaces with equal density in the ocean are called isopycns.

The oceans' thermal capacity is much larger than that of air: the top two meters of the ocean have a thermal capacity equivalent to that of the entire column of air above them. Because of weight and thermal inertia, oceans leave a very special mark on changes in the climate.

2.3.1. *Rotation, Coriolis, geostropy*

Like all points on Earth, the oceans spin east because of their rotation around the axis of the poles. Each point possesses an intrinsic speed of rotation to the east, which depends on its distance from the axis of the poles. This speed is maximal at the equator, then decreases progressively to cancel out at the poles. The existence of this rotation speed dependent on latitude affects the trajectory of fluid masses. Thus, a mass of water that moves from the North Pole toward the Equator will have a rotation speed to the east that is weaker than the surrounding masses: it will veer toward the west on its right relative to the surrounding milieu. Similarly, a mass of water moving from the Equator toward the North Pole, will have a rotation speed greater than the surrounding masses and will veer toward the east, to its right. The Earth's rotation imposes a spin toward the right on all movement occurring in the northern hemisphere and reciprocally, toward the left in the southern hemisphere. This spin causes a complementary acceleration, orthogonal to the trajectory, which is called Coriolis acceleration. The force that this engenders does not work and manifests itself only for a moving mass. We will designate the Coriolis parameter f, which characterizes the "planetary vorticity" at a given latitude: $f = 2\ \Omega \sin \phi$ where Ω is the Earth's angular speed and ϕ the latitude. This action of planetary rotation is determinant for most oceanic motions. Let us consider a reference point situated on the Earth's surface, where the axis Ox designates the longitude, Oy the latitude and Oz the local vertical. At this point, the planetary vorticity corresponds to the projection of the Earth's rotation on the local vertical.

In fact, most of the interior ocean is in a state called "geostrophic equilibrium". This equilibrium characterizes a stationary oceanic circulation, without acceleration or diffusion, where pressure forces that are exercised within the fluid are in equilibrium with the Coriolis acceleration:

$$f\,\mathbf{k} \times \mathbf{u}_h = -\,1/\rho\ \mathbf{grad}\, p$$

where $\mathbf{k}$ designates the unitary vector of the local vertical, $\mathbf{u}_h$ is the vector of horizontal speed in the latitude–longitude plane, with its two components u zonal speed and v meridional speed, ρ is the volumic mass and p the pressure. The equation above translates into the velocity components by:

$$-fv = -\,1/\rho\ \partial p \,/\, \partial x$$

$$fu = -\,1/\rho\ \partial p \,/\, \partial y$$

Schematically, when a zone of high pressure appears within the fluid, the water masses will diverge from this zone, and the Coriolis acceleration will lead them into a rotation to their right, i.e. rotation in a clockwise direction around this center of high pressure in the northern hemisphere, and the opposite in the southern hemisphere (Figure 2.7). This rotation movement around high pressures is called anticyclonic, whereas that which occurs around low pressures is cyclonic. Note: the cyclonic–anticyclonic rotation depends on the hemisphere where it is produced. The fluid motion tends to align itself with the isobars (the line of equal pressure).

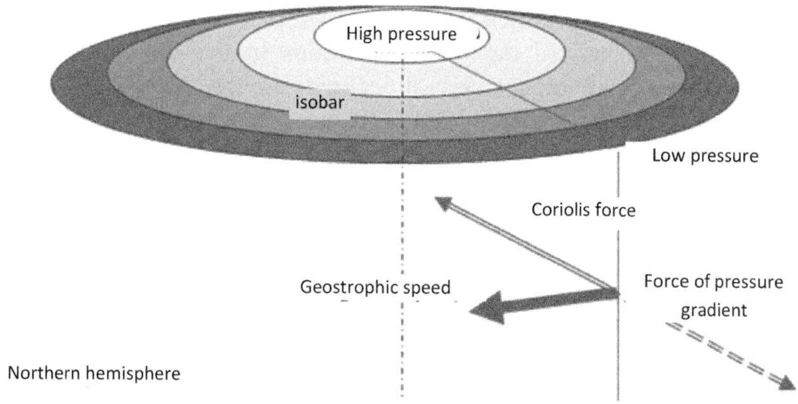

Figure 2.7. *Diagram representing the equilibrium between the Coriolis force and the force of horizontal pressure gradient, which creates a geostrophic current, parallel to the isobars. A case of anticyclonic circulation in the northern hemisphere*

2.3.2. *An ocean moved by the wind*

One of the most efficient motors for moving the ocean is the wind. In fact, when comparing a map of oceanic currents and a map of winds, there are a striking number of similarities between the two fields. Thus, the winds on the Atlantic Ocean are characterized by winds from the west at mid-latitudes, in the wake of which we find the Azores Current oriented east-south-east; further south, we find the trade winds, in the wake of which we see the North Equatorial Current, oriented toward the west. Oceanic circulation presents a large anticyclonic circulation at subtropical latitudes, whereas further north, in the subpolar region, we find a cyclonic cell. This situation, found in the North Atlantic, is also found in the North Pacific. In the southern hemisphere, we also note the large anticyclonic cells at subtropical latitudes, of which the southern branches combine to rejoin the Antarctic Circumpolar Current that links the three oceanic basins.

2.3.3. *Ekman, spiral, transport, pumping, upwelling and downwelling*

Very early on, oceanographers sought to explain oceanic circulation by surface winds and to understand the mechanisms by which the winds made their mark on the water mass. It was Ekman, in 1905, who rationalized the surface motions in the ocean following observation of the trajectories of icebergs in the Arctic. He noted indeed that the trajectory of the icebergs, whose center of mass is essentially underwater, did not follow the axis of the wind, but veered to the right of the windward direction. The wind drags the surface of the ocean by a force of stress, in the axis of the wind, which spreads by vertical diffusion in the layers beneath. In the interior ocean, the diffusion forces are in equilibrium with Coriolis acceleration, and when the full system of equations is resolved in an ocean with a constant vertical diffusion, the resulting motion is characterized by a spiral disappearing into the ocean while turning to the right, and decreasing exponentially toward the deep ocean. This is Ekman's spiral (Figure 2.8), which is the solution to the system of equations:

$$f\mathbf{k} \times \mathbf{u}_h = \partial \{A_v\, \partial \mathbf{u}_h/\partial z\} / \partial z$$

with, as a boundary condition at the surface $z = s$: $A_v\, \partial \mathbf{u}_h / \partial z = \boldsymbol{\tau} / \rho$, where A_v designates the coefficient of vertical diffusion, and $\boldsymbol{\tau}$ the wind stress vector at the surface.

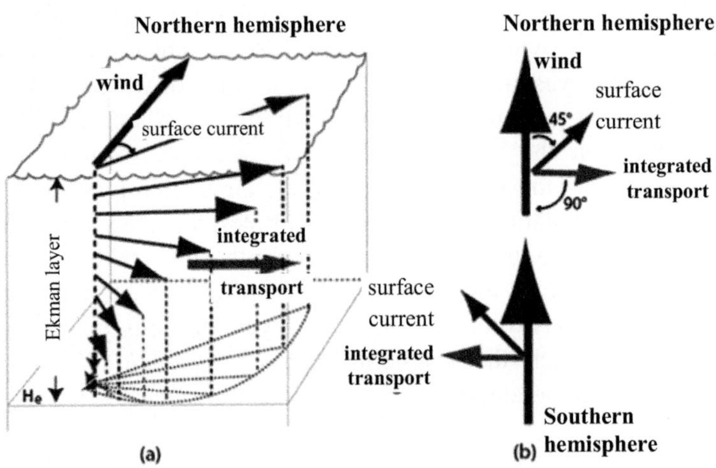

Figure 2.8. *a) Diagram illustrating the oceanic Ekman spiral in response to the wind stress at the surface; b) diagram illustrating the surface current and the Ekman transport integrated in each hemisphere (from [FIE 10])*

The surface current is at 45° to the right of the windward direction in the northern hemisphere, and at 45° to the left of the windward direction in the southern hemisphere. More than the surface current, we are often interested in the current integrated in the surface layer where the wind dragging effect is evident – this is Ekman's layer, a few tens of meters deep. The current within this layer is therefore at 90° of the windward direction and is called Ekman transport, perpendicular to the wind at surface. It characterizes the main motions of the surface layers of the ocean. Its expression is given by:

$$\mathbf{T}_E = \int_{\text{He->}s} \mathbf{u}_h \, dz$$

where $\mathbf{T}_E$ designates the Ekman transport, integrated between the base of the Ekman layer He to the surface s. The equation for the momentum in the Ekman layer is integrated in this layer:

$$\int_{\text{He->}s} f\,\mathbf{k} \times \mathbf{u}_h \, dz = \int_{\text{He->}s} \partial \, \{A_v \, \partial\mathbf{u}_h/\partial z\} \, /\partial z \, dz$$

From where:

$$f\,\mathbf{k} \times \mathbf{T}_E = \boldsymbol{\tau} / \rho_0$$

which allows us to calculate the Ekman transport: $\mathbf{T}_E = \boldsymbol{\tau} \times \mathbf{k} / \rho f$.

This expression of Ekman transport confers important properties on it: Ekman transport is produced in the surface layer of oceans, it is perpendicular to the windward direction, on the right in the northern hemisphere and on the left in the southern hemisphere, and its amplitude is inversely proportional to the Coriolis parameter f. Thus, for a given wind, the Ekman transport is even more substantial close to the Equator (Figure 2.8).

We will now focus on some particular properties resulting from spatial variations in Ekman transport. To illustrate the ocean's reaction, we will take account of the vertical structure of the ocean, with a relatively thin and light surface layer, separated from the interior ocean by a region very stratified in density or in temperature, with a strong gradient called the thermocline (we will return to this notion later in more detail). The surface ocean is reactive to the wind field, whereas the deep ocean remains motionless, the pressure forces there are null.

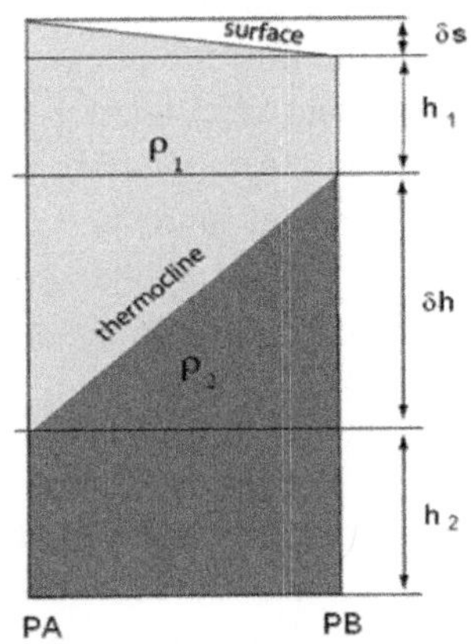

Figure 2.9. *Theoretical configuration of the two-layers of ocean*

This two-layer model (Figure 2.9) enables us to calculate the expression of the pressure forces in the deep layer and if we apply the hypothesis that the gradient of pressure forces cancels itself out there by equaling the pressure between PA and PB, we find:

$$\text{PA} = \rho_1\, g\, (\delta s + h_1 + \delta h) + \rho_2\, g\, h_2 = \text{PB} = \rho_1\, g\, h_1 + \rho_2\, g\, (h_2 + \delta h)$$

where ρ_1 is the volumic mass of the surface layer, ρ_2 that of the deep layer, δs is the surface elevation of the ocean and δh the elevation of the thermocline.

Thus:

$$\rho_1\, (\delta s + \delta h) += \rho_2\, \delta h$$

From whence δh is expressed:

$$\delta h = -\, \rho_{1/}(\rho_1 - \rho_2)\, \delta s$$

This expression shows that the surface slope and the thermocline slope are in opposition and in a ratio $\rho_{1/}\,(\rho_1 - \rho_2)$, which is in general very large since the difference in volumic mass at the interface is weak compared to the average volumic mass: the motions of the interface multiply the motions of the free surface a hundredfold, in the opposite direction. Consequently oceanographers adhered for a long time to describe the topography of the thermocline as a picture of the geostrophic circulation in the surface layer.

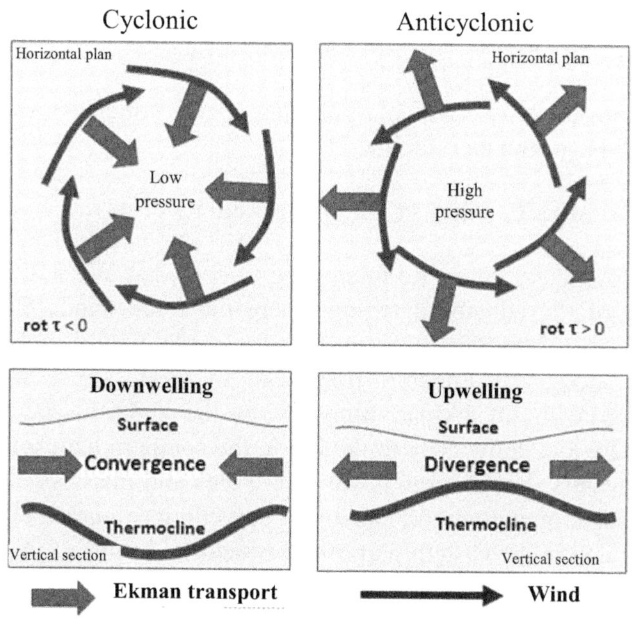

Figure 2.10. *Diagram illustrating Ekman pumping in the northern hemisphere (from [OUC 89])*

We will look at what is produced when an atmospheric low pressure passes over the ocean: the low pressure is marked by cyclonic winds curling around the center of low pressure in an anticlockwise direction (it is said that the rotation of these winds is positive). These winds drive an Ekman transport to their right, thus diverging: the waters of the surface layer are pulled toward the exterior of the low pressure, the surface level of the ocean sinks (the surface pressure diminishes), while the thermocline, between the surface layer and the interior ocean, rises (Figure 2.10). In the case of a low pressure, this rising of water from the bottom is called *upwelling* and we speak of oceanic *upwelling* due to Ekman pumping. The reverse instance, or *downwelling*, is produced under anticyclonic atmospheric circulation. This mechanism of Ekman pumping shows that the wind stress only acts on the surface layer but it will also impact the interior ocean through deformations in the thermocline that it sets in motion. The formalization of this equilibrium is calculated by integrating the continuity equation in the surface layer:

$$\int_{\text{He->}s} \mathbf{div}\ \mathbf{u}\, dz = \int_{\text{He->}s} \{\mathbf{div}_h\ \mathbf{u}_h + \partial w/\partial z\}\ dz = 0$$

We introduce w_E, Ekman's pumping speed, which is measured at the depth He of the thermocline:

$$w_E = \mathbf{div}_h\ \mathbf{T}_E = \mathbf{div}_h\ \{\boldsymbol{\tau} \text{ x } \mathbf{k} / \rho_0 f\} = \mathbf{curl}\ \{\ \boldsymbol{\tau} / \rho_0 f\}$$

Two particular cases are interesting to consider. We will first take the case of the equatorial region where the trade winds blow. It is possible to schematically assimilate the trade winds with a wind blowing along the Equator, from east to west. These winds are associated with an Ekman transport to the right in the northern hemisphere and conversely to the left in the southern hemisphere. The surface waters thus diverge toward the Poles and this loss of mass is compensated by a powerful equatorial upwelling (Figure 2.11). It is in this way that in the eastern part of the equatorial basins of the Atlantic and Pacific, we encounter surges of cold, deep water that are rich in nutritive salts.

The second particular case that we present concerns the example of the Senegal-Mauritanian Coast. The winds blow toward the south, along the coast and create a transport toward the west, pulling the surface waters toward the open sea. The departure of these waters creates a rise of water from the bottom along the coast, called coastal *upwelling* (Figure 2.12). The rise of cold water from the bottom also brings nutritive salts and favors the emergence of coastal productivity. The regions of *upwelling* have been identified for a long time as regions of intensive fishing and they are found along the subtropical coasts: the coast of Senegal-Mauritania, the coast of Namibia, the coast of Peru and the coast of California.

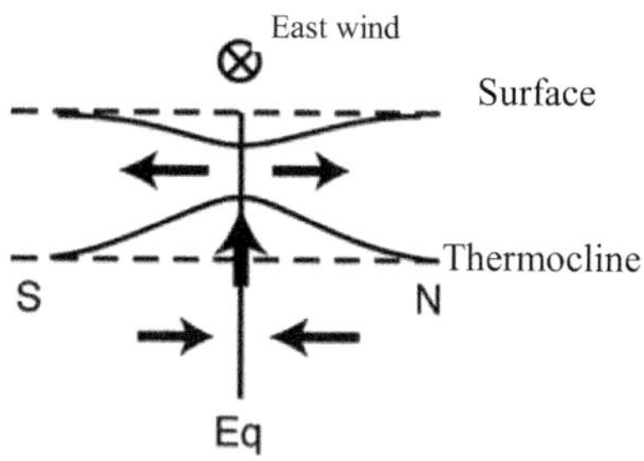

Figure 2.11. *Diagram representing equatorial upwelling*

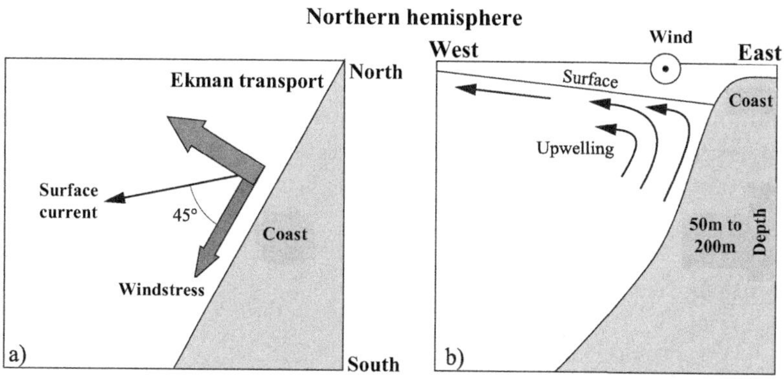

Figure 2.12. *Schematic representations of coastal upwelling in the horizontal plan 9 (a), in the vertical plan 9 (b) (from [FIE 10])*

2.3.4. *Interior ocean and western boundary currents*

Ekman's theory has proved a powerful concept for explaining the main characteristics of currents in surface layers in the central and eastern parts of the basins but there still remained an important question concerning the circulation in the interior of the basins and along the western coasts. One enigma persisted: what was the mechanism that could explain the strong currents along the western coast, such as the Gulf Stream? Indeed, neither the direction, nor the amplitude of such a current can be explained by studying local winds. It was therefore necessary to find another mechanism.

It was at the end of the 1940s that Sverdrup was able to understand the internal behavior of the ocean as he put forward a theory that would later carry his name. Sverdrup was interested in the circulation of the interior ocean, beneath the mixed layer. He considered a basin stretching from the tropics to mid-latitudes with, in its northern part, winds from the west and in its southern part, winds from the east, which corresponds to a negative vorticity, and causes a downward Ekman pumping, at the interface between the surface layer and the interior ocean. Taking account of the principle of conservation of the total vorticity on the sphere – the total vorticity is the sum of the relative vorticity of the fluid and the planetary vorticity f, this supply of negative vorticity imposes a southward displacement on the fluid, that is to say toward a latitude where the planetary vorticity f is weaker. He showed that in the wind field of negative vorticity, the fluid masses of the interior ocean move toward the Equator and conversely, in a wind field of positive vorticity, they move toward the Poles.

REMARK.– What is vorticity? It is the rotation capacity of a particle. Its formal expression is written as:

$$\zeta = \mathbf{curl}\ \mathbf{u_h} \cdot \mathbf{k} = \partial v / \partial x - \partial u / \partial y$$

The total vorticity is the sum:

$$\zeta_t = \zeta + f$$

One important consequence of Sverdrup's theory concerns the large western boundary currents (Figure 2.13). If, in an interior basin, the fluid mass moves southward, as the total mass is conserved at a given latitude, it is indeed necessary to find a northward movement to compensate and this is ensured by the western boundary currents such as the instance of the Gulf Stream, which rises in a longshore current, the equivalent of all the mass displaced southward in the entire interior basin. This western boundary current is made possible by the very energetic character of its interactions with the western shore, which engender the positive vorticity necessary for this displacement.

The southward mass transport in the North Atlantic Ocean, due to a wind, is in the order of 30 million m^3/s or 30 Sv (1 Sv or Sverdrup = 1 million m^3/s). As for the Gulf Stream, it does not transport 30 Sv northward but nearly 100 Sv. Its power is such that it has a self-pulling effect and creates a strong recirculation of the water masses, which increases its amplitude, with complex phenomena of nonlinear turbulent interactions. This phenomenon of amplified western boundary currents, rendered possible by the conservation of the total vorticity, which is also called the "β effect" where β characterizes the variation of the planetary vorticity according to the latitude, is present in the circulation of all basins, whether that is for surface currents or for bottom currents.

We have described here the instance of the intensification of western boundary currents in anticyclonic circulation at mid-latitudes; the same phenomenon is acting to produce western boundary currents in subpolar cyclonic circulations, or in deep circulation.

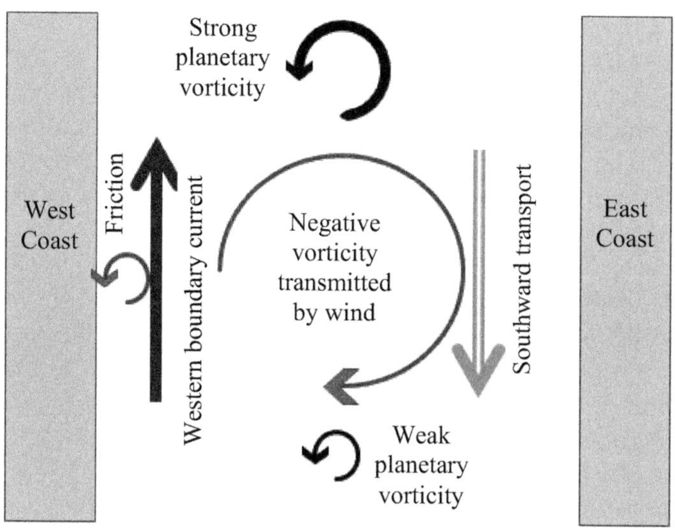

Figure 2.13. *Diagram illustrating the Sverdrup circulation and the western boundary current*

2.3.5. *An ocean moved by thermohaline fluxes*

We have just highlighted the importance of surface wind stress as a motor of oceanic circulation (and conversely, this wind stress at the surface of the ocean acts as a brake on atmospheric circulation). This exchange of mechanical energy is only one of the methods of exchange at the air–sea interface. Ocean and atmosphere also exchange heat, mass and numerous chemical components. The thermohaline fluxes correspond to surface exchanges that modify temperature and salinity.

Let us consider the surface heat budget. The ocean receives energy from the sun, of which it reflects a part; it also emits into the atmosphere in long waves, and it exchanges two sorts of thermal energy with the atmosphere: the sensible heat flux and the latent heat flux. The first flux depends on the difference in temperature between air and water as well as on the surface wind stress, the second flux depends on the specific humidity difference and on the wind stress. What interests us for the climate, is to understand the distribution of this surface budget and to analyze it in the context of the distribution in temperature in the oceans (Figure 2.14). The budget map shows that the maximum absorption of thermal energy in the oceans occurs in the central-eastern part of the equatorial oceans with close to 200 W/m^2 absorbed. The regions where the ocean loses a great deal of thermal energy to the atmosphere correspond to the large western boundary currents in the wake of which enormous quantities of energy are transmitted into the atmosphere. What we note here, is that these exchanges do not follow a regular distribution between Pole and equator, but they are strongly affected by the characteristics of the oceanic dynamics: the absorption of energy along the equator does not correspond to a maximum temperature; the presence of cold equatorial waters (the presence of *upwelling*) inhibits the release of latent and sensible heat in these regions, and favors a broad absorption of solar flux. In the west of the tropical basins, where the surface temperatures are between 28 and 30°C, the very significant latent heat flux tends to compensate the solar flux, which renders the gain very slight. In the large western boundary currents, it is the latent and sensible heat fluxes that very effectively extract the heat from the ocean to warm up the atmosphere above.

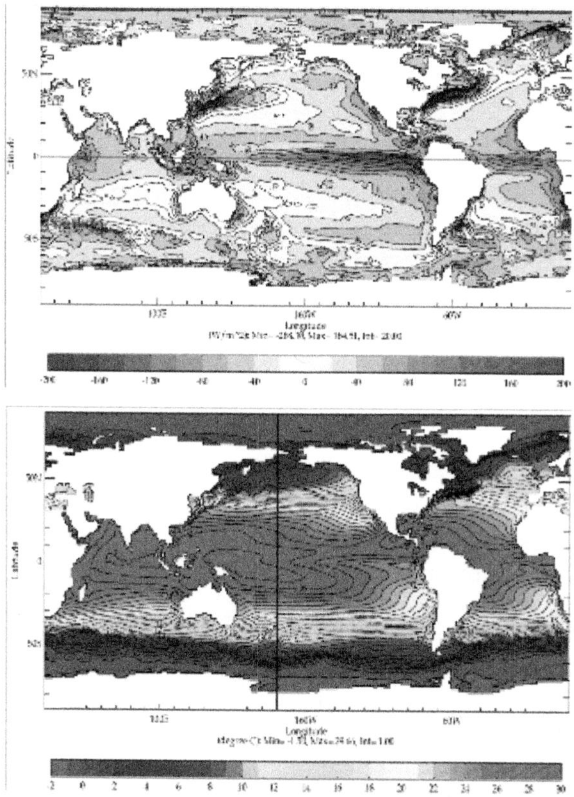

Figure 2.14. *Mean annual heat budget at the ocean–atmosphere interface from 1984 to 2007 in W/m² above and, below, mean annual temperature from 1984 to 2007 on the surface of the oceans in °C (data from http://oaflux.whoi.edu/heatflux.html) (© Sébastien Masson (LOCEAN/IPSL, CNRS/IRD/ UPMC)) (see color section)*

Ocean and atmosphere also exchange water at their interface. The ocean indeed gains mass from the *run off* from rivers and by atmospheric precipitations, it loses it by evaporation (which is the second interpretation of the latent heat flux). Globally, the ocean loses mass in the tropical belt; it gains it by the precipitations of west winds at mid-latitudes, and in the strong precipitations of the atmospheric intertropical convergence zones. These water fluxes are visible through variations in salinity: these increase in areas of evaporation and diminish in areas of precipitation.

2.3.6. *Stratification, mixed layer, thermocline*

The exchanges of surface fluxes determine the value of the surface's properties of temperature and salinity, and therefore the volumic mass. An initial approach is to consider the deep ocean as stable and to presume that only the surface ocean varies locally with the surface fluxes. In most parts of the oceans, the thermal fluxes are crucial for defining the stratification. At mid-latitudes, these fluxes have a marked seasonal cycle. They start to heat the ocean in spring, creating a fine and hot surface layer called the mixed layer. This layer becomes hotter with the advance of the summer season and deeper with the nocturnal mixing, when the surface heat budget becomes negative, and with the gusts of wind that create vertical turbulence. The region of significant thermal difference between the surface ocean and the deep ocean is called thermocline. It is close to the surface, between 50 and 200 m, and very sharp in tropical regions; it becomes deeper and wider toward the poles.

Diurnal and seasonal mixed layers form with the advance of the seasons, a positive surface heat budget contributing to stratify the surface layers, whereas the negative heat budget causes vertical mixing, which deepens the mixed layer. The main thermocline is defined as the most extreme position reached by the vertical mixing at the end of winter. This notion is important since the mean thermocline characterizes the limit of the ocean that will be in interaction with the atmosphere, whereas the ocean beneath remains isolated and only evolves in relation to the surrounding waters.

In mid- and tropical latitudes, the stratification is principally dominated by gradients in temperature, the salinity only becoming involved in regions of strong precipitation where episodic barrier layers form (the warm pool in the west Pacific, the Bay of Bengal or the mouth of Amazon). The notion of a "barrier layer" results from the fact that the supply of fresh water at the surface isolates the lower part of the mixed layer from the atmosphere, and thus creates a barrier effect for exchanges between the ocean and the atmosphere. In polar regions, the tough conditions at the surface tend to homogenize the temperature profile, and the variations in salinity are therefore responsible for the stratification.

2.3.7. *Formation of water masses, convection and subduction*

The subpolar regions are characterized by the formation of sea ice, which stars to form in autumn. When it forms, the freezing of the seawater throws excess of salt into the surrounding waters, the ice itself containing very little salt (4 g/kg). This excess of salt weights the surface waters, which sink deeper, to reach their level of density equilibrium. This phenomenon is called deep convection and it contributes to the renewal of deep oceanic waters. This convection is produced in an intermittent fashion, when the very cold ocean freezes at the surface, in the vicinity of the ice edge. It is however very important for the general circulation of the ocean, since it allows the formation of the North Atlantic Deep Water which flows between 2,000 and 4,000 m from the regions where it formed in the Norwegian Sea, in the Greenland Sea and in the Labrador Sea, into the South Atlantic. Regions of formations of dense water can be found in the open ocean (Figure 2.15(a)), as well as along the continental coasts, when wind blowing from the land pushes the ice into the open sea, thus exposing the ocean to a sudden cooling (Figure 2.15(b)). It is this type of phenomenon that occurs along the Antarctic, contributing to the formation of Antarctic Bottom Water. We note finally that this phenomenon of the formation of dense waters is not limited to polar regions. It can also occur when strong, cold, dry winds coming from the continent blow onto the sea, such as that which is produced in the Mediterranean. In fact, during successive gusts of the Mistral on the Gulf of Lion, the ocean does not freeze, but it becomes denser, under the effect of a flux of very intense evaporation (of several hundreds of W/m^2). The waters thus produced are deep Mediterranean waters, which flow through the Strait of Gibraltar and come to join the North Atlantic Deep Water, leaving their mark by a very salty trace, much hotter than the surrounding masses. Although it is produced in very limited areas, the phenomenon of convection is crucial for renewing all the masses of deep water and when water masses sink, they drag their surface properties down, including salinity and temperature and also their carbon concentration. It is in this way that a marked indication of the penetration of climate warming can be found in convection regions, as well as an increase in the concentration of anthropic carbon.

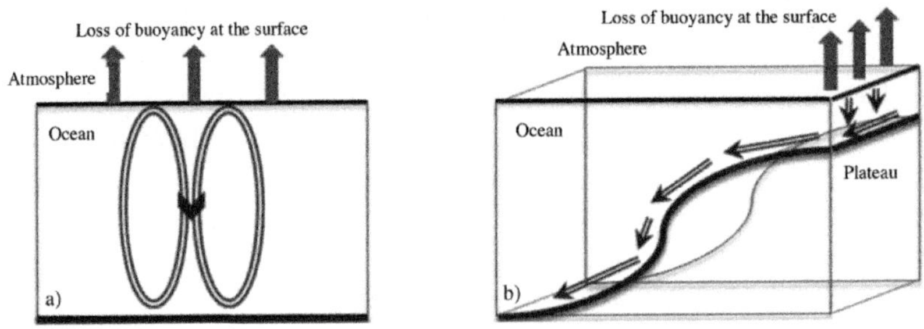

Figure 2.15. *a) Illustration of convection in the open ocean; b) illustration of convection along the edges of a continent*

The phenomenon of convection is not the only one capable of changing deep oceanic waters. These can also be altered by ventilation or subduction (Figure 2.16). This phenomenon occurs at the end of winter, or the beginning of spring, when the mixed layer reaches its maximal depth, in the main thermocline. This thick mass of homogenous water can slide into the thermocline waters, when the surface wind favors an Ekman pumping toward the bottom of the ocean (*downwelling*). A large part of the thermocline waters, called mode waters, is thus renewed.

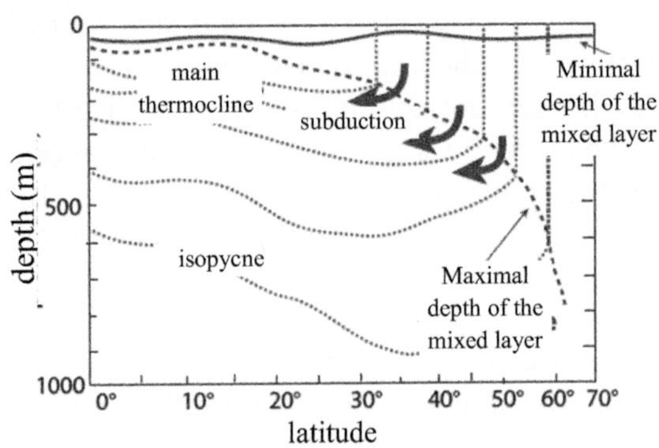

Figure 2.16. *Schematic representation of the phenomenon of subduction which enables the renewal of thermocline waters (from [FIE 10])*

The vertical distribution of temperature, zonally averaged for the global ocean, shows a general form in W, which is very characteristic.

In Figure 2.17, what can be found is the deepening of the thermocline at mid-latitudes and its presence close to the surface near the Equator, which makes the Equator very reactive to disturbances since the cold waters easily come to the surface. In the tropics, ocean and atmosphere are coupled and interact easily. Deeper regions do not have a direct link with the atmosphere and their water masses are renewed by advection from the surface through ventilation/subduction for the thermocline waters, and by convection for deep waters, then by diffusion and mixing with surrounding waters.

2.3.8. *Schematization of global circulation: the great conveyor belt*

General oceanic circulation results from the main mechanisms that we have just discussed: the formation of water masses at the surface by exchange with the atmosphere, injection into the depths by convection at high latitudes or subduction at subpolar/mid-latitudes, ascent of water through coastal and tropical *upwellings*, anticyclonic circulation in large mid-latitude basins and cyclonic circulation in polar regions, these thermal and dynamical phenomena being much more efficient at redistributing water masses and their properties than simple diffusion. The resulting image is complex because the geographical configurations of each ocean are different and each one of them develops a circulation unique to itself.

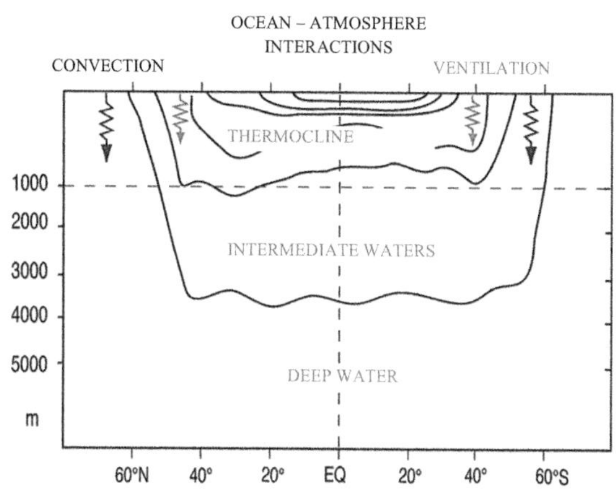

Figure 2.17. *Temperature profile of the global ocean, as a zonal average*

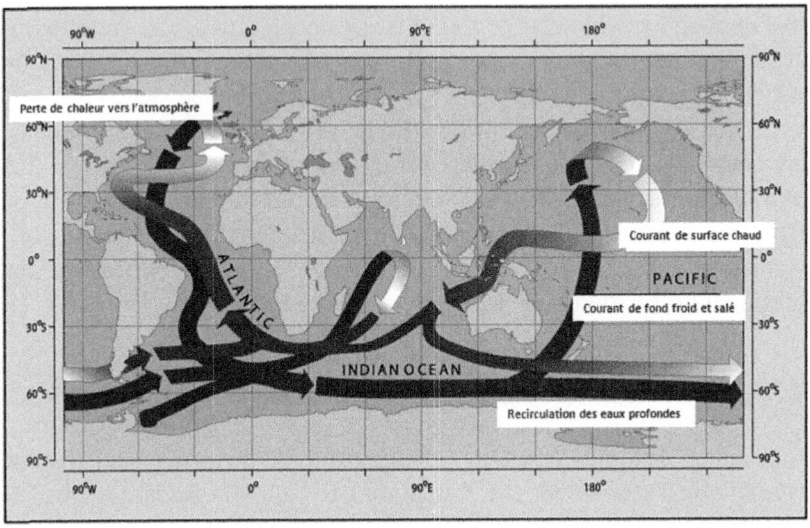

Figure 2.18. *Schematic representation of the general oceanic circulation conveyor belt (from [BRO 79], modified by E. Maier Raimer, AV/D3/99-2 (CLIVAR))*

A scheme allowing the general circulation to be illustrated has been suggested and popularized by Broecker [BRO 79] and remains attractive for understanding the global behavior of the ocean and the scales involved (Figure 2.18). It is effectively a *conveyor belt*, which enables the arrival of North Atlantic Deep Waters in the global ocean to be followed. We explain its main traits here:

– the North Atlantic Deep Waters leave the subpolar regions of the North Atlantic following the western boundary between 2,000 and 4,000 m; they are joined in the North Atlantic by Mediterranean Deep Waters;

– they cross the Equator and enter the Southern hemisphere, always on the western boundary which they only leave to be pulled into the Antarctic Circumpolar Current;

– around the Antarctic, they are intensely mixed and return progressively north in the thermocline waters in the Pacific, Atlantic and Indian basins;

– the branch rises toward the Equator in the Pacific, comes to supply the thermocline circulation, continues to warm and rises to the surface, describing several loops in the equatorial Pacific before reaching a depth level in the west, where it can cross the Indonesian Straits to enter the Indian Ocean;

– it crosses the Indian Ocean to reach the coasts of Madagascar and comes to supply the Agulhas Current to the south-east of Africa. The Agulhas Current is joined, in the Indian Ocean by a branch coming from the Antarctic Circumpolar Current;

– this water mass flows around Africa and enters the Atlantic Ocean in intermediate and surface layers by the Benguela Current and also in the form of large eddies; another branch also comes from the Antarctic Circumpolar Current in the South Atlantic;

– it reaches the Equator and after several loops of recirculation in its equatorial circulation, it crosses the Equator;

– it returns to the North Atlantic Ocean through the large western boundary currents, the North Brazilian Current, the Florida Current and then the Gulf Stream;

– it circulates several times in the anticyclonic Atlantic circulation before joining the North Atlantic Drift, then the Norwegian Current, which permits it to return to its region of origin.

This very schematic representation is useful for putting the communication between oceanic basins and the critical role of certain regions (the Indonesian Straits, the Cape of Good Hope, etc.) into relief, for showing the role of convection regions (the North Atlantic) and subduction (to the north of the Antarctic Circumpolar Current) and the ascents into the thermocline and into equatorial circulation, for contrasting residence times in equatorial regions (loops of some years duration), in temperate regions (loops of some decades duration) and in the rotation around the Antarctic, for showing the importance of western boundary currents and the transfer times associated with them.

Completing this long journey takes several hundred years, which is very rapid compared to typical diffusion times, but very slow compared to typical timescales in the atmosphere. It is thus possible to see here the importance of the oceanic dynamics, which by a tempo very specific to different regions, actively participate in keeping the planet's climate in equilibrium and participate in its variability.

2.4. Some key elements for understanding the ocean's role in the climate

2.4.1. *Typical times*

To understand the behavior of the climate system, it is necessary to contrast the different compartments that contribute to it and in particular to analyze the two large fluids that, through their continual motion, contribute to regulate the disparities in energy. The atmosphere is known for its rapid reactions, but also for its weak memory to disturbances in energy. A disturbance at mid-latitudes is transmitted in a few days into the surrounding regions and in less than a month to the entire hemisphere concerned. On a global scale, the Earth is in radiative equilibrium: all the energy that it receives from the Sun is emitted back into space. As a zonal average however, this energy equation is not balanced: the tropical regions receive energy in excess whereas the polar regions have a deficit. This planetary imbalance causes a redistribution of heat in which the ocean and atmosphere participate in equal parts (Figure 2.19). The atmosphere participates actively in the transfer of thermal energy from the Equator to the Poles by its average circulation in tropical regions (Hadley cell) and by its transient motions at mid-latitudes. Equally, the ocean participates in this transfer of energy from the Equator to the Poles notably by its northward heat transport in the subtropical basins but at its own speed. Moreover, in this meridional heat transport, each basin brings a particular contribution: the Pacific Ocean transports heat from the Equator to the Poles in a symmetrical manner; the Atlantic Ocean guarantees a heat transport to the north from the southern hemisphere, as far as the Arctic regions, whereas the Indian Ocean, bordered to the North by the Indian Continent, transports heat to the South.

The meridional heat transport in the ocean is produced by several mechanisms. For example, the anticyclonic circulation in the subtropics of the northern hemisphere carries along relatively cold waters southward and pulls warm waters northward by western boundary currents (wind-driven circulation). This results in a northward heat transfer. Another way to ensure this transport is the conveying of heat northwards by surface currents and drawing cold waters toward the Equator deep in the ocean: there is therefore an overturning loop that ensures this meridional transfer (thermohaline circulation). In the North Pacific, it is the anticyclonic circulation that ensures the poleward transport; in the North Atlantic, the two phenomena overlap and contribute to the northward transport. However, the typical timescales of these two phenomena are different: the balance of the wind-driven circulation depends at mid-latitudes on the propagation time of waves called Rossby Waves, which is in the order of a few years to a decade, whereas the overturning cells set in motion the transport of deep waters on a global scale, which occurs over hundreds to thousands of years through the occurrence of convection at high latitudes, which can be triggered very quickly.

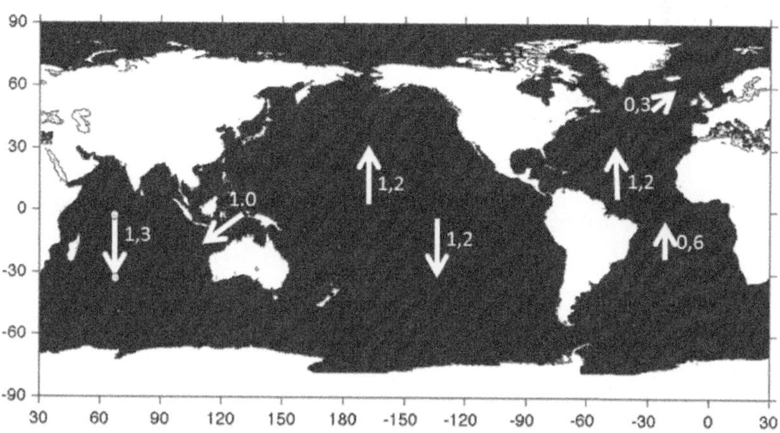

Figure 2.19. *Estimations of heat transport in different regions of the ocean. The transport is measured in Petawatts (which are 10^{15} Watt)*

2.4.2. *Ocean–atmosphere in the tropics*

In tropical regions, the ocean's dynamics respond rapidly to disturbances since the equatorial ocean, because of its sharp thermocline close to the surface and the cancellation of the Coriolis parameter, reacts as a wave guide [GIL 82]. The phenomena that keep the ocean in equilibrium with the atmosphere are not only advective but propagative and are measured in weeks or months. With this rapidity of reaction, the coupled evolution between the ocean and the atmosphere is thus easy: the ocean moves into a new state under the influence of the atmosphere, and this new ocean state causes the atmosphere to evolve. The two fluids evolve in concert.

If we seek to characterize the recurring patterns of variability in temperature at the surface of the oceans, once the seasonal variability has been filtered, a dominant mode is revealed, centered in the equatorial Pacific Ocean, with a broad spatial and temporal signature: it is El Niño. This mode is characterized by a pronounced warm anomaly in the central and eastern Pacific, which starts at the end of spring and continues and intensifies until the end of the following winter, until it disappears rapidly in spring. This anomaly in surface temperature is accompanied by a large modification in the oceanic and atmospheric circulations.

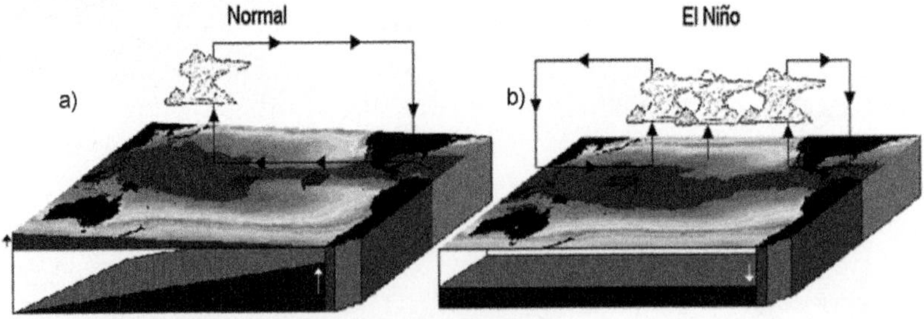

Figure 2.20. *Schematic representation of the ocean–atmosphere circulation in the Pacific. In a) its normal state and in b) its El Niño state*

First, let us describe the state of equilibrium of the atmospheric and oceanic circulations in the Pacific Ocean. The atmosphere is characterized by a constrated distribution of pressures in the tropics. Along a meridional section, the circulation is ensured by Hadley cells that, in the lower troposphere, draw masses of dry, hot air from tropical regions of high pressure toward the low pressure convergence zone, situated next to the Equator; these air masses, which gain humidity during their displacement, are then drawn at altitude by ascendance in the convergence zone and return at altitude towards the mid-latitudes, where they subside to low layers in the high tropical pressure regions situated either side of the Equator. Along a zonal section (Figure 2.20(a)), the circulation known as Walker circulation consists of a large loop of zonal circulation over the Pacific, marked by winds from the east in the lower troposphere, an ascendance of masses of hot, humid air over the West Pacific accompanied by heavy precipitation, then a return of air masses eastward, in the upper troposphere where the air dries and cools progressively, to return to the surface in the eastern Pacific. The winds from the east are supported by the difference in pressure between the eastern Pacific and the western Pacific. The ocean's response to these two cells is marked by an accumulation of hot surface waters in the west where the surface is higher than in the east by 60 cm, an important deepening of the equatorial thermocline (more than 200 m in the west), whereas the thermocline is close to the surface, around 50 m, in the east. The zonal circulation consists of a westward surface current (the South Equatorial Current) and an eastward current in subsurface (the Equatorial Undercurrent). Along a meridional section, the surface waters diverge from the Equator at the surface (divergence of the Ekman transport resulting from the trade winds) and converge in the subsurface. One peculiarity of the ocean is the complex organization of its circulation between two regions with singular properties: the geographic Equator, which is a region of geometric singularity for Ekman pumping, and the convergence zone of the winds, also called the meteorological Equator. The existence of the meteorological Equator is associated with a dipole of Ekman pumping (*upwelling* in the convergence region, *downwelling* between the convergence zone and the Equator) which gives rise to a very dynamic eastward geostrophic current: the North Equatorial Counter Current (Figure 2.21).

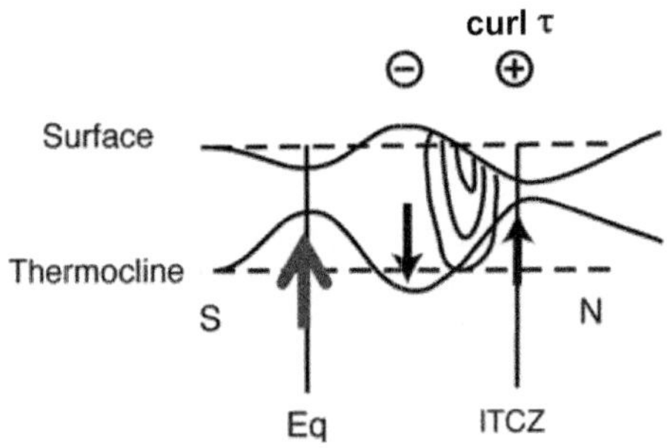

Figure 2.21. *Diagram illustrating the presence of the North Equatorial Counter Current, an eastward geostrophic current, between the Equator and the intertropical convergence zone (ITCZ)*

When El Niño occurs in the Pacific (Figure 2.20(b)), the ocean–atmosphere system reaches a state of equilibrium very different from the preceding state. The warm temperature anomaly at the surface in the central eastern Pacific diminishes the equatorial pressure gradient between the east and west of the Pacific, which slows down the Walker circulation. The Hadley cells are also affected; the intertropical convergence zones situated to the north and south of the Equator merge into a single region of convergence next to the Equator, in the central eastern Pacific, the location of strong rains. The pressure anomaly at sea level is negative in the high pressure region of the south-east Pacific and on the contrary is positive in the region to the north of Australia–Indonesia. The difference between these two centers of pressure is called the Southern Oscillation Index (SOI), which is recognized as the atmospheric signature of El Niño (Figure 2.22(a)).

The ocean is also subject to deep modifications: the occurrence of warm temperature anomalies in the central-eastern Pacific reduces the east-west temperature difference as much as the north-south temperature difference. The sea level evens out, as well as the thermocline, which corresponds to a redistribution of heat along the Equator and between the Equator and the convergence zones. The Equatorial Undercurrent thus diminishes, as the North Equatorial Counter Current, and a reversal can even be found in surface layers with an eastward surface current along the Equator at the culmination of the event. The Equator becomes a region of convergence rather than divergence of the surface waters. One of the direct consequences of El Niño is the disappearance of *upwellings*, which results in a spectacular impoverishment of the coastal and equatorial waters and ecological catastrophes, as much for marine life as for sea birds. The temperature anomaly in the equatorial Pacific (Niño region 3.4, between 5°N and 5°S, and 120°W to 170°W) is the El Niño index for oceanic anomaly (Figure 2.22(b)).

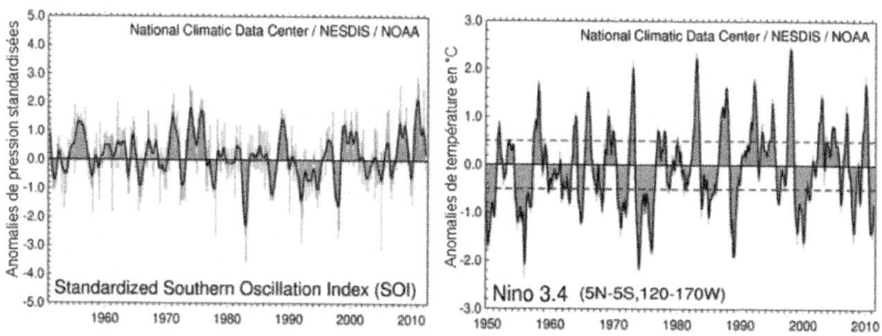

Figure 2.22. *Standardized Southern Oscillation Index, which characterizes the atmospheric signature of El Niño, on the left and on the right, the temperature anomaly index for the Niño region 3.4 of the Pacific, which characterizes the oceanic signature of the event (data available on www.ncdc.noaa.gov/teleconnections/enso/indicators/soi.php)*

El Niño is a frequent phenomenon, which repeats itself every two to ten years in the equatorial Pacific, with a favored frequency of three to five years. Its nature and origin remain subjects of debate: is it triggered by a local disturbance in the central Pacific or is its nature oscillatory? An initial disturbance of heat in the central Pacific translates rapidly into a growing coupled anomaly. The heat accumulated in the warmpool in the west Pacific favors the occurrence of gusts of wind from the west; some of them, close to the Equator, act on the ocean by propagating a pressure wave eastward, called a Kelvin wave, which deepens the thermocline in its wake, thus transmitting the hot anomaly eastward in one to two months. Whilst the temperature anomaly propagates, the surface winds slow down, which supports and even reinforces the initial anomaly; from whence comes the system's coupled amplification. This simple schema also contains the mechanism that will cause the end of the anomaly; indeed, the wind disturbance does not only trigger a pressure wave eastward, but also triggers pressure waves westward, called Rossby waves, which supply the equatorial high pressure anomaly by seeking water either side of the Equator, thus creating low pressure anomalies. These waves end by intercepting the west coasts of the Pacific where they reflect themselves more or less efficiently, to return in the form of Kelvin waves eastward, but this time the wave front that propagates east raises the thermocline, eroding the initial signal.

El Niño is a major phenomenon of climatic anomaly that affects most of the planet, in its meteorological and climate characteristics (precipitation in Peru and at the Equator, droughts in Indonesia and in Australia, cyclonic activity in Polynesia and consequences as far as North and South America during the culminating phase in winter) and whose persistence also affects the behavior of land and marine ecosystems. Its consequences are visible on a global scale: it affects the Earth's rotation, modifies the meridional heat transfers and the efficiency of greenhouse gas exchanges between the atmosphere and the marine and land biospheres. Its duration is around a year and it is sometimes followed by a cooling anomaly, which manifests itself by a reinforcement of the trade winds and the equatorial *upwelling*, called La Niña.

Despite the observation systems widely deployed for twenty years in the Pacific Ocean and progress in the study of this phenomenon, some questions are still unresolved: what are the causes of the beginning of an El Niño? What factors determine its amplitude? Will it be followed by a cold event (La Niña)? And above all is it influenced by or participating in the evolution of the Earth's climate, along with climate change? Two major events occurred at the end of the last century and are imprinted in our minds: the event of 1982–1983 and that of 1997–1998. El Niño plays a regulating role in energy transfers on an almost global scale but we only have a limited history of it and the coupled models used to study the climate and its evolution can barely capture all its characteristics.

The other equatorial oceans are not exempt from variability. The Atlantic Ocean shows an anomaly very similar to El Niño at the Equator, but it has a much weaker amplitude and scarcely lasts a season. However, it also shows a degree of variability over decades, characterized by a 'seesawing' of the centers of pressure between the southern hemisphere and the northern hemisphere. This large-scale anomaly modulates the interhemispheric transfer through the trade winds as well as the oceanic heat transport.

As for the Indian Ocean, it is marked by an anomaly called the zonal south Indian Ocean Dipole. It is characterized by a warming in the region to the north of Madagascar and a cooling to the west of Australia. This oceanic situation accompanies winds from the east on the Equator, which reverse the equatorial surface current toward the west whereas in this ocean, the equilibrium situation is characterized by the presence of eastward jets along the Equator (or Wyrtki jets), well marked in the intermonsoon periods. This situation is associated with a deficit in precipitation to the east of the Indian Ocean and reinforcement to the west. This raises the question of whether this anomaly, which presents as a mirror anomaly of El Niño, is effectively linked to it. The answer is not simple since the Indian Ocean was one of the first to warm and to modify its variability; it seems that the links between its own variability and other external sources such as El Niño have changed over the last century.

2.4.3. *Other types of variability*

Climate change poses a challenge for its detection in ocean and atmospheric environments. The long (and still very rare) observation series that we have available show indeed that the climate is altering a lot, with spatial and temporal characteristics that have been defined progressively, though it is difficult to know whether they are a natural manifestation or a sign of change, since the signals are interlinked and our time perspective too short. Our knowledge of modes of variability has increased spectacularly during the last decades. Numerous types of oscillation have been identified and characterized by their signature in terms of anomaly patterns in surface temperature such as the North Atlantic Oscillation, the Arctic Oscillation, the Atlantic Multidecadal Oscillation, the Pacific Decadal Oscillation, the Southern Annual Mode, etc. Each of these modes affects the ocean but is this a question of noise, of transitory disturbances or of a durable impact exercised by the atmosphere on the ocean? And can the persistence of certain modes be explained by a feedback action of the ocean on the atmosphere?

This last hypothesis in particular has been studied for North Atlantic Oscillation (NAO) in the North Atlantic. The NAO is characterized by an anomaly in pressure difference between Iceland and the Azores. Its positive phase corresponds to a decrease of low pressure for Iceland and a reinforcement of high pressure for the Azores, which accelerates the flux that crosses the Atlantic to the east toward northern Europe. Therefore, there is a disturbed situation with precipitation in the north of Europe and on the contrary dry conditions in the south. The negative phase corresponds to an anomaly of inverse pressure, so to a decrease of the pressure gradient and a weakening of the flux from the west, which is oriented more zonally: it rains in the south of Europe. Contrary to El Niño, there is no dominant frequency in the NAO spectrum. However, we note decadal periods where a particular phase of the anomaly dominates [HUR 95]. The persistence of the NAO has been explored by several groups who have examined how an initial hot anomaly in temperature off Florida could find itself buried in the thermocline over winter to surge back, in the axis of the Gulf Stream the following year, contributing to the reactivation of the atmospheric anomaly, but in a slightly different region. This oceanic resilience, which can explain the multiannual persistence of the atmospheric anomaly, warns us to be careful regarding the evolution of climate change: what is buried in the ocean finishes by returning to the surface, with the ocean's characteristic tempo, but also its inexorable force.

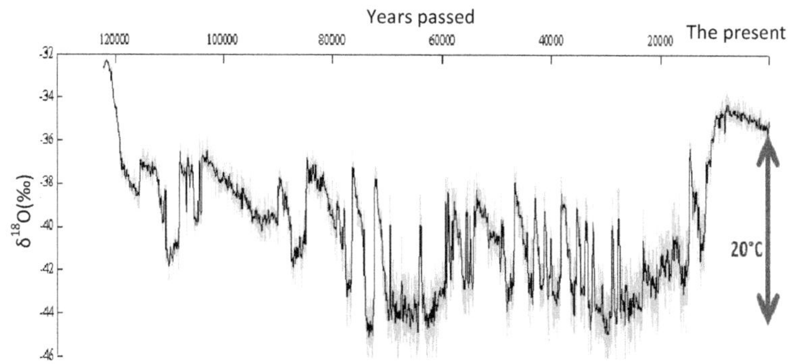

Figure 2.23. *Data for the isotopic composition of ice in oxygen 18 in a core from the site NorthGrip, in Greenland (from [NOR 04])*

2.4.4. *Climatic surprises*

In the course of understanding the climate variability and the associated risks, paleoclimatologists invite us to better regard and understand what has already occurred on the Earth in other epochs, particularly during the last million years. The last Ice Age, which ended around 8,000 years ago, is very informative. Records from ice cores from Greenland allow us to examine the oxygen isotope content contained in air bubbles trapped in the ice as they were transformed from snow into ice. This variable is conveyed in air temperature. The ice core, associated with a model of the ice age, enables us to reconstruct the history of the atmospheric local temperature up to the last million years in the northern caps, such as that of Greenland, and for the entire period of glacial–interglacial cycles in the Antarctic caps. The first thing that strikes the observer in the information from Greenland (see Figure 2.23) is the modification in the shape of the curve over time: the temperature of the last 8,000 years is almost flat, whereas the preceding glacial period shows dramatic variations of more than 10°C. The world of the Ice Age was a world of large climatic variability, whose instability was without doubt due to the large quantity of polar caps, these being very unstable. The oceanic data confirm the existence of instability during the last Glacial Maximum. They show sedimentary data interspersed with sequences of mineral debris drawn from continents, signs of the propagation of icebergs very far south. These events, called Heinrich events, indicate that oceanic circulation has without doubt experienced important modifications and suggests that the cause of the instability is linked to thermohaline circulation; this phenomenon of instability in thermohaline circulation, well marked in the data from the last Glacial Maximum, is it susceptible to being reproduced today, in the context of a climatic warming that might destabilize the current caps?

Theoretical studies have focused on control phenomena for the formation of deep water in particular regarding the surface exchanges that fix the extremes of temperature and salinity which lead, or not, to convection. The two variables, salinity and temperature, contribute to the phenomenon of convection: the temperature is fixed by exchanges of heat fluxes between the ocean and the atmosphere and in return it acts strongly on the flux itself (negative feedback on the fluxes of latent heat and sensible heat); the salinity depends on the sum of "evaporation minus precipitation" whose value does not depend, or depends very little, on the salinity: there is no feedback between the fields. This differentiation in the formulation of forcing has important consequences on the dynamic system that flows from it. The fact that the oceanic stratification depends on two variables – one controlled, the other not controlled – introduces a solution with chaotic character: a single flux, "evaporation minus precipitation" can lead to different solutions for the oceanic convection zones. The current situation is that of an active North Atlantic, but it could turn to an active Pacific, with important consequences in terms of heat transport in the northern hemisphere. The phenomenon that has been called "thermohaline catastrophe" following the work of Stommel corresponds to a cessation in the formation of deep water in the North Atlantic. This phenomenon has without doubt happened in a glacial climate, following an extension of sea ice over the formation zones of deep water, thus intensifying the evolution of very cold conditions, but could this occur in the near future?

The observations and projections for the future climate are scrutinized to discover how the Gulf Stream would evolve as well the northward heat and mass transports. The observations are still too sparse to reconstruct a long history. Data from recent years show fluctuations in the Gulf Stream but do not enable us to learn much about its evolution. The radials made between Greenland and Iceland show evolution in the properties of deep waters, which are becoming warmer. But there too, the data is still insufficient. In projections for the next 100 years created with coupled models, the tendency is for a slowing down of the meridional transport associated with the overturning circulation, but it should be noted that the current ocean models scarcely give this transport's magnitude, there is therefore still some way to go.

Sensitivity studies have been carried out to test the ocean's behavior in centennial climate projections. In particular, the rate at which the ice melts is an important factor in the efficiency of oceanic convection. The water issued from the melting ice is fresh and flows at the surface, isolating the interior ocean from interactions with the atmosphere. This thin surface layer prevents the phenomenon of convection. Imposed on simple dynamic systems, this phenomenon leads rapidly to a cessation of thermohaline circulation, which leads to a cooling of northern regions. This sensitivity has been tested in the framework of a coupled general circulation model with which two climate change experiments have been carried out in parallel. These two experiments were constructed from the current situation as far as a doubling in the concentration of CO_2, a concentration that then remains constant for several hundred years. The only difference between these two experiments came from the imposed melting rate of the ice on the Greenland cap. The first experiment had no melting rate; whereas for the second, a uniform melting rate was imposed. The difference between these two experiments, at the end of 500 years, is represented on Figure 2.24(b). It can be noted that the imposition of a melting rate leads to a very significant cooling of the climate at high latitudes, as much as 10°C in certain regions of northern Europe. The oceanic convection zone, situated in this model to the east of Greenland, shows a slight warming. These events indicate that the thermohaline circulation slowed down a great deal and a more precise diagnosis shows in fact that the cooling in this model is due to feedbacks between clouds and atmosphere in the sea ice boundary zone, more than to modification in the thermohaline circulation. Moreover, when we look at the field of temperature anomalies compared to the model for the present (see Figure 2.24(a)), we can state that the global anomaly remains positive over Europe. Therefore, by imposing a melting rate, this model shows much strong feedback between the ice and the ocean–atmosphere climate system with a significant lessening of warming in Europe, but it does not reverse the warming tendency, as some other dynamic models suggest.

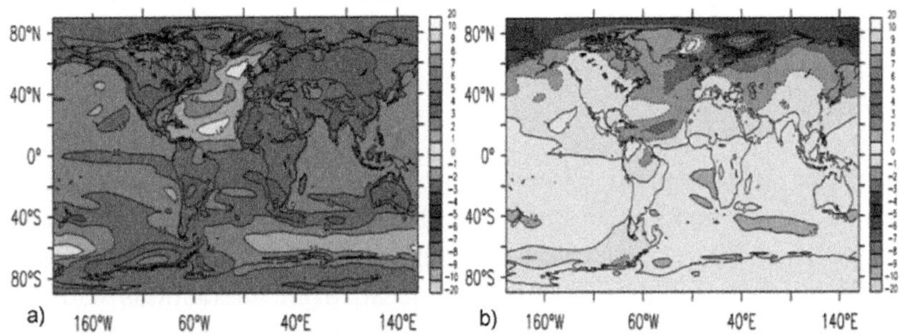

Figure 2.24. *Anomalies in annual global temperature calculated by the coupled ocean-atmosphere model in a simulation in which the concentration of CO_2 is doubled with an imposed melting rate for ice. Map of the difference after 500 years with the temperature of the model for the present state 9 a), map of the difference after 500 years with a coupled simulation without ice melting 9 b) (© Didier Swingedouw (LSCE/IPSL, CEA/CNRS/ UVSQ)) (see color section)*

2.5. Some questions for the future

Observations of temperature, salinity, and carbon show that the ocean is in the process of slowly accumulating anomalies linked to modifications in the climate. On a global scale, the ocean is warming, with signs that are well marked in the first 700 m but difficult to detect at depth. The penetration of surface disturbances at depth depends on phenomena that renew the waters of the interior ocean: thus the depth reached by the anomalies in warming and anthropic carbon is maximal in the subpolar regions of the North Atlantic. In the same way, regions of subduction are also particularly important for accelerating the penetration of signals in the deep ocean. The oceanic observations however remain difficult to interpret: very few long oceanic series, for several decades, exist; this length of time is necessary to construct a chronology long enough to distinguish a tendency toward warming from natural variability, which has a very rich spectrum, notably on seasonal to decadal scales. This variability masks the anthropic signal as their signatures are too interlinked to allow for detecting changes and attributing their causes. Moreover, the temperature and the salinity are not passive tracers, but they fully participate in the oceanic dynamics, by acting on the pressure forces. In a warming climate, with a more significant warming at the Poles than at low latitudes, the meridional heat transfer should slow down; it is expected that the overturning circulation will diminish too. This tendency has been searched for in oceanic projections produced with coupled general ocean–atmosphere circulation models and it seems to be effectively confirmed with time. But the observations are still far too insufficient for a conclusion to be drawn concerning the dynamics, notably in the North Atlantic. Repeated sections are programmed at different latitudes to pursue this monitoring, they show warmer waters at the bottom in the subpolar region, but do not allow conclusions on the evolution of mass and heat transports in the North Atlantic, taking account of the large spatial and temporal variability in the transports along the western boundaries of the oceans. In the Mediterranean, the climate warming should lead to a decrease in the formation of deep waters and a modification of their properties, which seems to be observed in this basin considered, as a critical zone of change on account of the significant risk of general drought in its perimeter.

Today, the ocean absorbs nearly half the anthropic carbon emissions, through solubility and by absorption in the biosphere. The solubility depends on the surface temperature: a hot ocean will absorb less carbon than a cold ocean. The absorption by the surface biosphere will lead to a carbon export from surface layers into the deep ocean through detritus and the deaths of organisms. The anthropic carbon, buried in the depths, will supply sediments but a part will return to the surface, in a few hundreds or thousands of years, according to the efficacy of the global circulation; from where comes the importance of detecting the modifications underway at the surface as well as in subsurface, and of understanding the reasons.

One particularly studied region is the Austral Ocean. For several decades, and in particular since the 1970s, the Southern Annular Mode has been in a positive phase, which tends to increase the zonal wind around the Antarctic. This increase in wind leads to an increase in northward Ekman transport, which could contribute to accelerating the rise of deep waters rich in carbon, and therefore to the saturation of the absorption capacity of anthropic carbon into the cold surface waters of the Austral Ocean, which would be positive feedback of the climate system, accelerating the warming. Indeed, recent studies show that the dynamics of this region are not that simple. The intensification of the average circulation is in fact compensated by the circulation of eddies, which are very active in the Antarctic circumpolar current. This leads to modifications in regions of absorption but also to convection and ventilation, with, moreover an important meridional variability.

The debate on the evolution of the absorption capacity of the ocean is still largely open and it highlights the importance of the interactions that are manifested as much among the physical and biochemical phenomena as among the processes at different scales. Our knowledge of the ocean is recent: the last few years have brought about as many new observations through the ARGO floats and through satellites as the entire 20th Century. This difficulty of reconstructing a long-term history of the ocean is a major obstacle to our understanding of its equilibrium on a large scale in an Earth system whose trajectory is not in equilibrium today.

2.6. Bibliography

[BRO 79] BROECKER W.S., "The great ocean conveyor", *Oceanography*, vol. 4, pp. 79–89, 1979.

[CAR 85] CARISSIMO B.C., OORT A.H., VONDER HAAR T.H., "Estimating the meridional energy transports in the atmosphere and the ocean", *Journal of Physical Oceanography*, vol. 15, pp. 82–91, 1985.

[FIE 10] FIEUX M., *L'océan planétaire*, illustrations by ANDRIE C., Presses de l'ENSTA, Paris, 2010.

[GIL 82] GILL A.E., *Atmosphere–Ocean Dynamics*, Academic Press, London International Geophysics Series, vol. 30, 1982.

[HUR 95] HURRELL J.W., "Decadal trends in the North Atlantic Oscillation: regional temperatures and precipitation", *Science*, 39, vol. 269, pp. 676–679, 1995.

[MOS 08] MOSS R.H., BABIKER M., BRINKMAN S., *et al.*, "Towards new scenarios for analysis of emissions, climate change, impacts and response strategies", *Intergovernmental Panel on Climate Change*, Geneva, Switzerland, 2008.

[NOR 04] NORTH GRIP-COMMUNITY-MEMBERS, "High resolution climate record of the northern hemisphere reaching into last interglacial period", *Nature*, vol. 431, pp. 147–151, 2004.

[OUC 89] OPEN UNIVERSITY COURSE TEAM, *Ocean Circulation*, Pergamon, Oxford, Open University, Milton Keynes, 1989.

[PHI 90] PHILANDER S.G.H., "El Niño, La Niña and the Southern Oscillation", *International Geophysics Series*, Academic Press, San Diego, vol. 46, 1990.

[SOL 07] SOLOMON S., QIN D., MANNING M., *et al.* (ed.), "Climate change 2007: the physical science basis", *Contribution of Working Group I to the Fourth Assessment Report of the Intergovernmental Panel on Climate Change*, IPCC, Cambridge University Press, Cambridge and New York, 2007.

[SVE 42] SVERDRUP H.U., JOHNSON M.W., FLEMING R.U., *The Oceans: Their Physics, Chemistry and General Biology*, Prentice-Hall, Englewood Cliffs, 1942.

[SWI 07] SWINGEDOUW D., BRACONNOT P., DELECLUSE P., *et al.*, "The impact of global freshwater forcing on the thermohaline circulation: adjustment of North Atlantic convection sites in a CGCM", *Climate Dynamics*, vol. 28, pp. 291–305, 2007.

[TRE 92] TRENBERTH K.E., *Climate system modeling*, Cambridge University Press, Cambridge, 1992.

3

Ocean–Atmosphere Interactions

3.1. Introduction: what are ocean–atmosphere interactions?

Ocean–atmosphere interactions involve exchanges of heat, mass (water) and chemical components. They bring mechanisms at different scales into play, from thermal conduction in water and air to coupling phenomena involving the entire atmosphere and upper layers of the ocean, such as tropical cyclones.

The essential source of energy for the Earth is the Sun. The Sun's radiation is centered in the visible domain, which reaches the Earth's surface after penetrating the atmosphere. The sea surface reflects some of this radiation, while the remaining radiation penetrates the ocean surface. The light can penetrate several tens of meters the ocean surface, allowing a transfer of energy into the upper layers of the sea. This heat supplied to the ocean can be transported by currents or redistributed into the atmosphere by means of heat exchanges or evaporation, which supplies the atmosphere with water vapor. The presence of surface inhomogeneities, such as oceanic fronts (variations in surface temperature over a few tens of kilometers) or sea ice, leads to horizontal variations in ocean–atmosphere exchanges.

Wind friction, as well as the transfer of heat between ocean and atmosphere, is strongly amplified and accelerated by atmospheric turbulence on the smallest of scales. Vertical displacements of air or

---

Chapter written by Laurence EYMARD and Gilles REVERDIN.

water parcels are linked to horizontal movements (wind and current) and to a vertical temperature or humidity gradient. In parallel, the wind friction at the surface triggers local phenomena (waves, or even sea spray and bubbles when the waves break), which themselves influence exchanges of energy and matter between the water and air.

Moreover, the ocean and the atmosphere exchange matter: water (by evaporation and precipitation), gases and solid particles. Gas exchanges depend on the wind speed and water temperature (through the gases' solubility). The deposited particles are carried, through the atmosphere, from the continents by dry deposition or dissolved inside rain drops; the emitted particles result from the evaporation of sea-water from sea spray ("primary" aerosols), and also, more indirectly, from the chemical transformation of gases issuing from organic components of surface water (DiMethyle Sulfure (DMS) or organic volatile components), in particular inside clouds.

First, we will summarize the physics of air–sea exchanges of energy and matter. Then we will show, using a few examples, how these processes intervene in the ocean–atmosphere coupling on different scales.

3.2. Interface processes and their role in the coupled system

In this section, we will summarize the bases of the processes involved and their formulation. We will show how exchanges of energy and matter are involved in the ocean–atmosphere coupling and then we will give the main approaches to measure them.

A surface energy flux is the quantity of energy that crosses a unit of the surface in 1 second. It is expressed in Watts per meter squared. The energy balance at the surface can be written as:

$$G = Rnet - Fturb \tag{3.1}$$

with:

– G: flux penetrating the ocean;

– *Rnet*: net radiative balance;

– *Fturb*: turbulent heat fluxes.

3.2.1. *Radiative fluxes*

The Sun's spectrum covers the domain of electromagnetic waves, ranging from ultraviolet waves to microwaves, including visible and infrared bands.

Like all bodies with a physical temperature, the wavelength distribution of radiation emitted by the Sun is a function of its temperature. Planck's law expresses this relation for a black body: a body whose surface entirely absorbs the incident radiation.

Because of Planck's law, the maximum energy radiated by the Sun is situated in the visible domain, approximately 0.5 μm, whereas the Earth, whose surface temperature is in the order of 290 K, has a maximum emission in thermal infrared (10–12 μm).

The atmosphere is not totally transparent to solar radiation: the gas molecules, the droplets, cloud crystals and dust particles (aerosols) have a significant absorption in infrared bands. They also reflect and scatter the received radiation (mainly in the visible spectrum, as well as in infrared and microwaves in specific frequency bands). Thus, the atmosphere absorbs part of the solar radiation, and it emits in turn a thermal infrared radiation downward and upward; the clouds reflect part of the radiation upward, and so into space; they also absorb, re-emit and scatter the remaining in all directions.

On the surface, the radiation received is the sum of all these components: solar radiation attenuated and diffused by the atmosphere, and infrared radiation emitted by the atmosphere. Strictly speaking, it would be necessary to consider the transmissions and emissions in the entire spectrum. However, the balance of energy received by the Earth's surface is largely dominated by short waves (visible spectral domain, from 0.4 μm up to near infrared, approximately 3 μm) and infrared (from 3 to 12 μm).

3.2.1.1. *Solar radiation*

The net solar radiation (short waves) is the sum of the incident solar radiation diffused by the gases and particles, in the atmosphere, minus the reflected radiation, which is function of the surface albedo. This albedo (reflecting power) ranges between 0 and 1, from a black body to a perfectly reflective mirror. Snow, sea ice and sand have high albedos (in the order of 0.8), whereas water has a low albedo (approximately 0.1), so that visible light, above all at short wavelengths, penetrates into the upper layers.

On average, the solar flux value is 1,365 Wm^2 before it enters the atmosphere on a disk perpendicular to the radiation direction. But because the Earth is round, the flux received effectively depends on the latitude. It also varies with the season. Diffusion through the atmosphere (which gives it its blue color) leads to a significant loss of this radiation, since some of the solar flux is sent back into space. Clouds also diffuse and reflect a more or less significant part of the solar radiation into space.

3.2.1.2. *Infrared radiation*

Net infrared radiation (long waves) denotes the difference between the radiation emitted downward by the atmosphere's components and the Sun, and the emission from the surface, which depends on its emissivity (between 0, a perfectly reflecting body, and 1, a black body). The sea has a high emissivity so it is an emitter of infrared radiation.

The spectral absorption of the atmosphere is a function of its components. Absorption in the atmosphere is mainly due to water vapor, carbon dioxide, methane, ozone, and numerous trace gases that are present in small quantities in the atmosphere. For any gas, concentration determines its absorption in one or more specific spectral bands. Water vapor is the most abundant component, and its spectral absorption domain is also very broad (many spectral bands). The clouds also absorb infrared radiation at all wavelengths.

This absorption (and therefore emission, according to the law "any body in thermodynamic equilibrium that absorbs radiation re-emits the same quantity") by the atmospheric components is responsible for the greenhouse effect: the presence of some cloud types, and above all of water vapor and CO_2, prevents the cooling of the lower atmosphere and surface. For example, on a clear, dry night, the atmosphere contributes very little to warming the Earth's surface, which thus cools by emitting infrared radiation.

3.2.1.3. *The radiation penetrating the ocean*

The radiation penetrating the ocean is partly used up through phytoplanktonic activity and through absorption in phytoplanktonic cells, enabling photosynthesis produced by certain pigments, but in particular chlorophyll pigments. In compensation, these cells, along with particular organic matter called colored matter (often in a dissolved state), re-emit the radiation in less energetic wavelengths (by fluorescence). However, the proportion stored in the form of chemical energy is weak, and the energy absorbed is rapidly transformed into heat, leading to a warming of the layers. These cells, just like other marine particles of biological or mineral origin and numerous dissolved substances, also refract part of the light received. From these interactions results a greater vertical "thermal" stratification of the surface layers. Moreover, light penetrates less deeply when these particles are present, which inhibits photosynthetic production in the deepest layers. Because of the surface warming, the long wave fluxes emitted at the surface of the ocean are significantly increased. Furthermore, the presence of organic and inorganic particles and dissolved colored matter in the surface layers slightly modifies the short wave radiation re-emitted by the surface (an albedo effect); these two effects have an impact on the planet's radiative balance.

3.2.2. *Turbulent fluxes*

While the surface radiative fluxes depend on numerous "players" outside the interface (notably, sunshine and clouds), the turbulent fluxes result from direct, local interactions between a moving atmosphere and the sea, that combine the lower limit of the atmosphere and a player in the interaction, through the movement and transformation of the air and water involved.

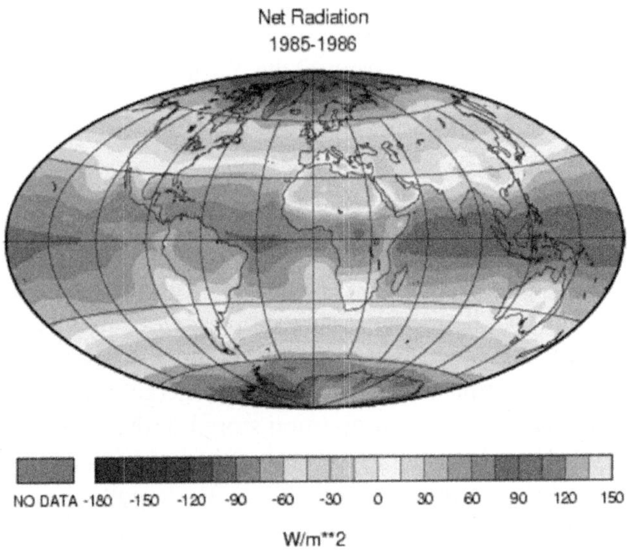

Figure 3.1. *Net annual mean radiative flux (data from the satellite ERBE, http://eos.atmos.washington.edu/cgi-bin/erbe/disp.pl?net.ann.) (see color section)*

In the atmosphere, the layer that interacts directly with the surface is called the atmospheric boundary layer (ABL). Its thickness, above the oceans, varies from several tens of meters to a maximum of 1,500–2,000 m. The lowest part of this layer, called the surface layer (a few tens of meters, less than 10 m over sea ice; estimated to be approximately 10% of the height of the ABL), is the site of turbulent exchanges, due to its contact with the surface. Strong gradients of temperature, humidity and wind take place in this layer. Above, the ABL can display a positive, neutral or negative stability: the transfers of heat from bottom to top are easy, possible, or the transfer can only occur downward, respectively. At its top, a transition layer separates it from the free atmosphere, in which large-scale meteorological systems take place.

Similarly, a fine surface layer is also found in the ocean (of a depth ranging from less than 1 m to several meters). It constitutes the top of the oceanic mixed layer (OML), equivalent to the ABL, where transfers occur between the surface, with its waves, bubbles and turbulent structures, to the OML. The OML is the layer where water in contact with the atmosphere mixes with the cooler layer below. As in the atmosphere, a stratified transition zone called the thermocline separates the OML from the deeper layers. This layer is described more precisely in section 3.2.3.

The ABL is characterized by its temperature, humidity and wind profiles. The OML is characterised by temperature, salinity and water currents.

The exchanges between the atmospheric surface layer and upper ocean (the first few centimeters) are governed by friction, due to the wind and currents, and by the differences in temperature and the humidity between them. The thermal conduction, density and thermal inertia of the atmosphere and ocean are such that turbulent movements of the atmosphere predominate in the calculation of the turbulent fluxes: the atmosphere responds almost instantly to variations in surface temperature, which drives the heat fluxes, and surface currents are, on average, of an order of magnitude lower than the wind speed.

3.2.2.1. *Surface fluxes: above all a problem of atmospheric turbulence*

Turbulent fluxes are expressed as the product of the atmosphere density (assumed to be locally constant), by the covariance of the vertical speed fluctuations, w' and of the horizontal wind u', temperature T', humidity q', depending on the calculated flux:

– momentum flux:

$$\tau = \overline{\rho}.\overline{u'w'} \qquad [3.2]$$

– sensible heat flux (C_p is the air thermal capacity):

$$H = \overline{\rho}.C_p.\overline{w'T'} \qquad [3.3]$$

– latent heat flux (evaporation flux E times the latent heat of evaporation L):

$$L.E = \overline{\rho}.L.\overline{w'q'} \quad [3.4]$$

These equations convey the fact that the atmosphere, through its turbulent movements, transports air vertically, along with its properties. The larger and more positive the difference in temperature between the surface and lower atmosphere, the more free convection is observed, with a significant heat flux. Similarly, a difference in humidity favors evaporation.

Horizontal wind plays two roles in the turbulent fluxes: it acts on the friction directly, thereby on the transmission of energy to the ocean and atmosphere. But a strong wind also favors heat fluxes by increasing the turbulence. Considering the major role of horizontal wind, approximate formulations have been developed which are easier to manipulate than the covariances: the flux parameterizations.

The direct determination of the fluxes requires the variables (temperature, humidity, horizontal and vertical wind) in the atmospheric surface layer to be measured with a very accurate measuring tool, and a high acquisition rate (up to at least 10–20 Hz). Indeed, this measurement is difficult to implement over the sea surface, since such instruments cannot be installed on isolated buoys. Moreover, these types of instrument have only existed since the 1980s.

Since the middle of the 20th Century, this technical difficulty has led researchers to look for formulations that can relate fluxes to mean variables, which are easier to measure.

Prandtl (1925) defined the "friction velocity" as equal to the square root of the momentum flux, divided by the density. Considering that turbulent eddies progressively lose their coherence, he defined a "mixing length", l_m, such that the friction speed is equal to l_m times the vertical gradient of horizontal speed. Using this, it can be shown that wind increases from the surface according to a logarithmic law, by linking l_m to a characteristic height called the roughness height (z_0):

$$\bar{u} = \frac{u_*}{\kappa} \log \frac{z}{z_0} \qquad [3.5]$$

By establishing empirical laws to determine z0, it becomes possible to link the fluxes to the average wind.

Using similar dimensional relations, the characteristic heights for temperature and humidity z_T and z_q, as well as T_* and q_* have been defined such that the covariances are replaced by the product $u_*.T_*$ and $u_*.q_*$ in the equations for heat fluxes. Monin–Obukhov's theory develops these concepts, permitting the calculation of flux in any stability condition. The methods developed since the 1980s for calculating flux, called global aerodynamics or "bulk", are based upon Monin–Obukhov's work. Fluxes are expressed as simple functions of the mean differences in wind, temperature and humidity between the surface and measurement altitudes. This approach requires the introduction of an exchange coefficient (C_D, C_H, C_E) for each flux, which is a function of the roughness height (or the typical magnitude u_*, T_* and q_*):

$$\tau = \bar{\rho}.C_D.(\overline{U - Us})^2 \qquad [3.6]$$

$$H = \bar{\rho}.C_P.C_H.(\overline{U - Us}).(\overline{T - Ts}) \qquad [3.7]$$

$$L.E = \bar{\rho}.L.C_E.(\overline{U - Us}).(\overline{q - qs}) \qquad [3.8]$$

This approach is the most commonly used, since it allows the estimation of fluxes using "ordinary" measurements, such as those made on moored buoys. Fluxes can be iteratively calculated. Mean quantities are measured over several tens of minutes, at a height of two meters to several tens of meters, then algorithms developed by various authors can be used to calculate the exchange coefficients (see, for example, the analysis of [BLA 87]). This is also the method used to calculate the fluxes in numerical weather models or for climate simulations. But, this simplicity has led to widespread use, sometimes disregarding that these formulae are only valid at small scales in time and space (in general, less than an hour and less than 50 km) which correspond to the scale of turbulent eddies in the atmosphere.

3.2.2.2. *Turbulent fluxes and the ocean*

For the upper layers of the ocean, the turbulent fluxes are the defining factors:

– the heat fluxes (marked here positively for an upward transfer of heat, in accordance with meteorological convention) cool the surface by temperature transfer and evaporation. These effects are added to the radiative balance to determine the proportion of heat transferred to the ocean (see Figure 3.2);

– the momentum flux, which conveys the friction exercised by the atmosphere on the sea's surface, has several effects:

- transfer of momentum, causing a displacement of surface water;
- turbulent mixing in the upper layers;
- generation of waves that then spread deeper;
- formation of swell (sea swell from the wind).

The shortest waves contain fairly little energy and rapidly retransmit their quantity of movement to the ocean when they break, which can cause turbulence in the ocean's upper layers in combination with the movement of orbital waves (or Langmuir cells). The longest waves are organized in the form of swells, whose energy propagates fairly rapidly from the regions where they were formed to the coastal zones, where they can sometimes trigger (especially in combination with a local wind) strong, breaking waves and the phenomenon of storm surges.

The quantity of momentum transported by the turbulence in the oceanic boundary layer (often bounded at its base by a stably stratified layer, the thermocline) will allow the creation of currents on a larger scale. Apart from the Equator, these currents are constrained by the Coriolis force linked to the Earth's rotation. Because of their horizontal structure (the same as that of the winds) they can be associated with horizontal convergences or divergences, and so cause strong horizontal pressure forces. This is the main source of the ocean's dynamic on a large scale.

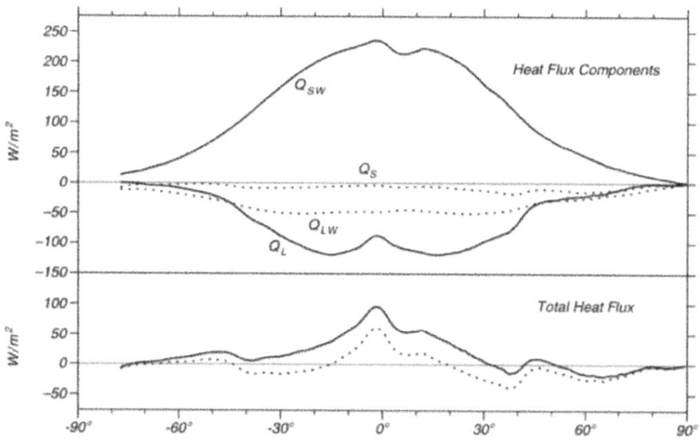

Figure 3.2. *Terms of the energy balance at the sea's surface, in zonal average (from [DA 94])*

COMMENTARY ON FIGURE 3.2.– The incident solar energy (marked as Q_{SW} on this diagram) is at its maximum in the tropics. The surface temperature being the highest in this latitude belt, the long wave fluxes (Q_{LW}), sensible heat (Q_S) and latent heat (Q_L) are also at their highest there. It can be observed from the net balance (flux entering the ocean, bottom diagram) that the sum of these terms displays a marked maximum near the Equator (precipitations in the equatorial belt greatly exceed evaporation).

3.2.3. *Water exchanges between the ocean and atmosphere*

We have seen that evaporation contributes to the energy balance at the surface. However, there is also a transfer of matter involved, contributing, as with precipitations, to the water balance at the surface:

– in the atmosphere, evaporation from the sea surface is a component of the hydrological cycle, where more than 80% of evaporation occurs over the oceans. Let us remember that water, in its different states, strongly contributes to the Earth's radiative balance, in its vapor stage as a powerful greenhouse gas and through the screening effect of clouds (see section 3.2.1). This water is transported over the continents by winds; after precipitation in watersheds it returns to the oceans by flowing in rivers or underground;

– in the ocean, evaporation is also the direct cause of salinity changes, and therefore of the ocean's surface density. It thereby acts on the local vertical circulation, or on a large scale due to the thermohaline circulation. Regions with excess precipitation can be strongly stratified by this supply of fresh, less dense water. A less dense surface layer can isolate the ocean beneath from the mixing caused by exchanges at the surface. As they receive light, the layers below can be warmed by the penetrating solar radiation, so that they become as warm as, or warmer than, the surface. In this case, they are called 'barrier layers'. The impact of such layers on the dynamic of *El Niño Southern Oscillation* (ENSO) events in the equatorial Pacific Ocean has often been mentioned (these layers are often present in the west of this basin) where they contribute to the maintenance high temperatures, even after wind bursts. The presence of such layers can also lead to amplification of tropical cyclones (for example, in the Bay of Bengal and the tropical Atlantic to the east of the Caribbean Sea);

Water is still involved in the ocean–atmosphere exchanges in its solid form, through the formation of sea ice. Sea ice develops during the polar winter, because of the sea surface cooling (through evaporation and radiative cooling). During the formation of ice, the sea salt is progressively pushed down toward the ice bottom in the form of brine, which increases the seawater's density underneath. This very dense, cold water can sink to the ocean bottom. Partially submerged, the ice layer is in contact with the atmosphere and ocean, and exchanges energy with both.

Sea ice formation processes are complex. The first crystals appear at the surface at around –2°C in the form of needles, accompanied by brine drops. Progressively, the ice crystals agglomerate, forming a frozen "soup", a viscous layer moved by the sea, of depths from 1 or 2 cm to 1 m: the frazil. At the same time, brine droplets agglomerate and sink. The floe forms progressively (several weeks are needed for a 60 cm thickness to be reached), little-by-little, pushing out some the bubbles of air and the pockets of brine.

The ice becomes translucent and blue. The different stages of formation have been given informal names in local languages: grease, slush, pancakes, hummocks, floes, etc.

Sea ice is important in the surface energy balance:

– on the atmospheric side, the main ice characteristic (covered or not by a snow layer) is its high albedo: solar flux is strongly reflected, and so barely contributes to ocean warming. Moreover, the ice sublimation process does not lead to a significant increase of atmospheric water vapor, contrary to the evaporation of water, in polar climatic conditions. On the ice, the energy balance is nearly reduced to the radiative balance. During the polar night, almost nothing but the net long wave radiation remains. Note that the polar atmosphere is characterized by a very thin atmospheric boundary layer, at the bottom of a thin troposphere (less than 10 km), in which clouds of ice very often occupy the entire depth. The radiative balance is strongly constrained by this cloud cover. However, as soon as the sea ice fissures, evaporation dominates the energy balance, contributing further to the generation of fogs and mists;

– on the oceanic side, the brine pockets differ strongly from the overall OML characteristics; the OML is not very salty at these latitudes (30–35 per thousand). Moreover, the heat and fresh water fluxes are not significant at the ice base once this has been formed. However, the ice formation is associated with a significant release of brine, which increases the density of the surface water under the ice. These fluxes can be enough, particularly in certain regions beside the continental plateau, to cause the formation of dense waters that can sink into the deep ocean.

Another interaction among the ocean, sea ice and atmosphere, is the impact of the wind and currents on the ice fields: displaced by friction, the ice floes fragment or collide with each another, creating temporary channels of free water between plates and polynyas (close to the continents, these are "lakes" of free water created by the action of the wind blowing from the continent). These free water areas are characterized, contrary to the ice, by a flux of intense evaporation, since the polar atmosphere is very dry (because of the low air temperature).

The balance of energy in sea ice regions is, therefore, very different depending on the frozen surface.

3.2.4. *Other exchanges of matter*

The other exchanges of matter between the two environments, ocean and atmosphere, are numerous and extremely important for the climate and the dynamic and biogeochemical behavior of the oceans and atmosphere. The gaseous exchanges of carbon, in the form of carbonic gas (CO_2), constitute an important sink for this gas with a strong greenhouse effect (approximately 48% of anthropogenic production has been absorbed by the ocean since the beginning of the industrial era, even if this proportion has tended to diminish), and therefore the exchanges of carbon have had a regulatory effect on the evolution of the climate.

The exchanges of matter that will be discussed, therefore, concern compounds dissolved in precipitations, exchanges of more or less soluble gas (in both directions) and solid particles (in both directions).

Particles, gases or dissolved compounds coming from the atmosphere can interact chemically (or photo-chemically) in seawater and eventually can cause biological production, particularly in euphotic layers by photosynthesis or, if they are organic in nature, they cause bacterial production that will break them down. They can also have a pollutant role (for example, certain organic compounds or certain “toxic” mineral compounds).

Some gases, resulting from chemical reactions linked to biological activity then emitted into the atmosphere, play an important role in the atmosphere’s radiative balance (for example, the oxide of di-nitrogen N_2O, or the sulfur of dimethyl or DMS $(CH_3)_2S$, without, of course, counting carbon dioxide.

The particles deposited on the sea ice modify its albedo (for example, black carbon) or undergo heterogeneous chemical reactions (in aqueous environments) at these interfaces. Those particles coming from the surface can act as nuclei for condensation in clouds (salts coming from sea spray and particulate sulfates coming, for example, from oxidation reactions of DMS), and indeed influence the atmosphere's oxidizing power (halogen or bromide compounds, N_20, etc.). The presence of sea ice inhibits or strongly reduces a significant proportion of the exchanges between the ocean and atmosphere, even if the deposits on the ice (and the frozen water in the ice as well as the snow) eventually rejoin the ocean. Gas emissions, in particular, are strongly reduced by the presence of sea ice.

Let us return to these different modes of ocean–atmosphere exchange by evaluating their main roles and in some cases identifying the processes that are responsible for these exchanges.

3.2.4.1. *Exchanges of gas*

These exchanges are understood in a context very similar to that of evaporation, but while noting that the equilibrium occurs between a dissolved phase of the gas and a vapor phase at the liquid–air interface (a free air–sea surface or on the walls of bubbles), or during evaporation from spray. For each gas, there exists a linear solubility ratio between the dissolved concentrations and partial pressure in water (equal to that in the air when in equilibrium) of which the coefficient depends in the first place on the temperature and salinity (Henry's law). This solubility is weak for certain gases, such as oxygen, but strong for others, such as carbonic gas. This partial pressure is linked to the atmospheric pressure, possibly modified for the behaviors of non-perfect gases, while taking account of humidity for a saturated atmosphere at the temperature of the air–sea interface.

The specific character of gas exchanges is that the profile in water of a dissolved gas displays a significant gradient beside the interface, as a result of molecular diffusion. In practice it is the weakest diffusion, the diffusion in water, that controls the state at the interface. This enables the partial pressure in air of the gas concerned to be estimated using measurements taken at a certain distance from the surface, without the local exchanges having a noticeable effect on this atmospheric composition; this is especially the case with oxygen, for which the measurements are greatly facilitated. We also note that the molecular layer concerned in the water has a slight thickness, a thickness (and gradient) that is expressed as a function of Schmidt's number $Sc=\nu/\varepsilon$ (or ν is the cinematic viscosity and ε is the molecular diffusivity of the gas).

The law of exchange, which is written as:

$$F=-K([A_a]-[A_w]) \qquad [3.9]$$

where $[A_w]$ and $[A_a]$ are the concentrations in water (beneath the diffusive layer) and air, has been the subject of numerous studies in the course of the past 30 years, (see, for example, [LIS 86, WAN 92]), which resulted in important difference on global estimations of exchanges of gas, notably for carbonic gas. This coefficient, K, is expressed as a function of Schmidt number (to an appropriate power that depends on the state of the interface: a stable film for very weak winds, or by renewal of the interface in a regime with intermediate wind) and a friction speed in water u_{*w}, linked to the friction speed of the wind at the surface. The exchanges at the level of bubbles play a particularly important role for gases that are slightly soluble, such as O_2, and can explain a slight oversaturation in oxygen at oceanic surface layers (from 1 to 2%). Laws of exchange dependant on the variability of sea surface slopes have also been proposed, which appear to take good account of the effects of surface waves and surfactants. The latter are more or less hydrophobic films, leading to tensio-active modifications of the surface, which influence the spectrum of short waves as much as the exchanges of gas. These laws of exchange are less used, however, because data for surface slopes are less readily available than those for the wind.

3.2.4.1.1. Surface concentrations of dissolved gas

For most gases, in particular inert gases, concentrations in water are always fairly close to concentrations in the atmosphere, despite seasonal variations in the saturation solubility in the ocean (change in temperature, as well as changes in atmospheric pressure). This results from the rapidity of equilibration in mixed oceanic layers due to exchanges with the atmosphere (particularly true for very volatile inert gases, such as helium).

This is less true for certain gases with a higher Schmidt number, or in the presence of very thick layers, for example, during deep convection events, which, over a limited time, place an important water reservoir often far from saturation in contact with the atmosphere. This is even less true for a gas such as DMS, which is produced during plankton blooms in the ocean and with weak atmospheric concentrations, because of its rapid chemical alteration in the atmosphere, and in particular its transfer into the clouds. For other gases that are stable in the atmosphere, such as O_2, N_2O and CO_2, the surface concentrations can be far removed from the equilibrium, for reasons that we will illustrate below in more detail. For example, concentrations in N_2O far removed from the equilibrium are found in narrow regions, linked to the remineralization of organic matter, which causes a significant air–sea flux of N_2O. These regions contribute to approximately 40% of global sources of N_2O in the atmosphere.

3.2.4.1.2. Oxygen

For oxygen, the oversaturation of oceanic surface layers increases during periods of warming and when phytoplankton production occurs, with typical oversaturation values of 3–10% during these episodes, lasting over periods of several weeks. However, in a large majority of oceanic regions, surface concentrations are relatively close to saturation. Concentrations at saturation being weaker at higher temperatures, the warming of surface waters over the course of the last 50 years has resulted in a reduction in surface concentration of oxygen in the sea. This development, perhaps coupled with a reduction in the ventilation of waters at the subsurface, due to an increase in stratification and in some cases to a change in the winds, has led to a reduction in concentration of oxygen beneath the surface in these regions and in numerous other oceanic regions. This phenomenon is particularly marked off tropical regions of *upwelling* (zones where water rises from deep layers) and in eastern boundary currents. In these regions, the phenomenon is magnified, the oxygen being consumed at the subsurface by bacterial activity, which enables the oxidation of organic matter. These regions of reduced oxygen, virtually anoxic, have tended to increase greatly in area over the course of he last 40 years, by at least 10–20%, leading to important consequences for the biogeochemical functioning of these systems, as well as for the food chains and ecology of these regions.

3.2.4.1.3. Carbon dioxide

In the context of gases dissolved in the surface layers which are close to saturation, the distribution of carbon dioxide gas in these surface layers is abnormal, with some notable differences saturation (of 10–30%) with respect to extensive oceanic regions. Equatorial and tropical latitudes tend to be source regions, whereas the high latitudes tend to be sink regions, although these sinks have been weakened greatly over the course of the last 30 years [LE 03], despite CO_2 being very soluble in seawater. This remarkable property comes from the fact that carbon dioxide is not an inert gas in seawater, and that it is associated in acid-base reactions that involve two other forms dissolved in seawater: carbonates and hydrocarbonates. At the ocean's typical pH of around 8.1, these two components represent 99.5% of the stock of inorganic carbon in seawater, whereas CO_2, in a hydrated form, (H_2CO_3), only represents 0.5% of this stock. As a result, the majority of inorganic carbon is found in the oceans, and only a small portion in the atmosphere. It takes a very long time, from a season to a year, to balance the oceanic surface layer of carbonic gas with the atmosphere above.

Moreover, CO_2 plays a major role in biological activity, when, during photosynthesis, it forms organic matter or inorganic skeletons (calcite and aragonite), as well as during respiration and the degradation of organic matter through bacterial activity. These processes are controlled, on the one hand, by the thermal impact on the partial pressure of surface waters (pCO_2), due to seasonal warming (which tends to increase the pCO_2 in the surface, and therefore to cause oversaturation during these warmings) and, on the other hand, by the impact of biological activity. Phytoplankton blooms contribute to carbon sinks near the water surface. In addition, the exported matter leads to carbon dioxide input into the deeper layers through remineralisation and respiration. Moreover, oceanic circulation, horizontal as well as vertical (or the vertical mixing of masses of water), will modify these surface signals which would tend to be associated with significant vertical gradients.

On the surface, the winter season is associated with an increase of inorganic carbon, which leads to an increase of pCO_2, more or less compensated by a decrease caused by the cooling. These different effects present different equilibriums depending on the regions, thicknesses of the mixed layer and biological regimes. Over the course of the last decades, these effects have been strongly modified by the great increase in atmospheric concentrations of carbon-based gases. We also note that the penetration of anthropogenic CO_2 in the oceanic surface layers has, due to the acid-base anthropogenic, triggered a decrease in the pH of the oceans. This decrease, called acidification of the oceans, which has already reached more than 0.1, can have significant effects on the solubility of the carbonated skeletons of certain marine species, coral reefs or carbonated sediments.

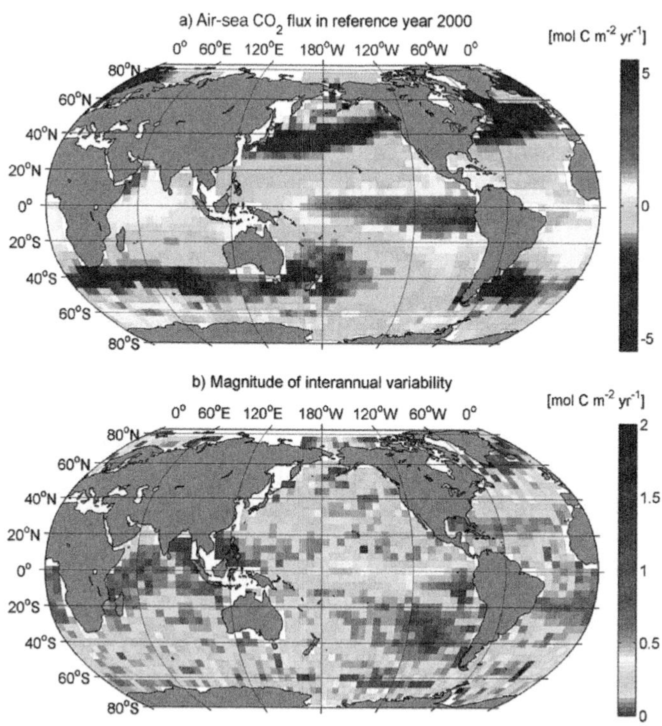

Figure 3.3. *Map illustrating the average air–sea flux of CO_2 at the interface (diagram above). Positive (negative) values correspond to a flux into the atmosphere (into the ocean) (from the site www.pmel.noaa.gov/) (see color section)*

COMMENTARY ON FIGURE 3.3.– These maps reflect the disequilibrium between the partial pressures in water and air. The map on the bottom illustrates the interannual variability of these fluxes. It is particularly strong in the equatorial Pacific Ocean, due to the ENSO phenomenon. The subpolar gyre of the North Atlantic also presents a strong decennial variability.

3.2.4.2. *Supplies of precipitation*

Rain is foremost a source of fresh water (see section 3.2.3), which generates a decrease in the salinity of surface water, and therefore an increase in its vertical stratification. In certain tropical regions, it acts to inhibit the vertical mixing (for example, barrier layers in the west Pacific). Rain is also a source of gas or dissolved particles, for example, in the case of solid hydrophilic aerosols that are transported into the atmosphere. It is also a source of dissolved salts, in particular, of inorganic nitrogen in the form of nitrates. The latter have increased over the course of the last decades, to a non-negligible degree in certain regions, whereby we can expect modifications at the level of primary production and the biogeochemical cycles in oligotrophic regions (those poor in nutritive salts). Other salts and important elements are carried by rain with divergent effects: from iron and other metals, to micronutrients for the growth of plankton whose availability is often limited, but also inorganic or organic micropolluants.

3.2.4.3. *Solid deposits*

Other additions to the ocean occur in solid form: for example, aerosols of Saharan origin, or organic particles, such as black carbon. In certain regions, these supplies could be an important source of the nutrients or micronutrients useful for the growth of phytoplankton, notably in the North Atlantic subtropical gyre or in the Northwest Pacific. Elsewhere, they can, however, act as a ballast and pull "useful" particles downward from the surface layers. Understanding the respective part of the supplies to the ocean in solid or dissolved forms remains an active subject of investigation, as well as the evaluation of their impacts on the oceanic surface layers and their biogeochemistry.

3.2.4.4. *Export of particles*

Finally, the ocean's surface layers are also the source of primary aerosols during the evaporation of sea spray, or from the bursting of bubbles at the surface. They mediate the export of different salts, which often contribute to the condensation nuclei of clouds, thus modifying their radiative properties and concentration in the regions concerned. These primary aerosols can also lead to secondary reactions in clouds and have impacts on the atmosphere's oxidizing properties. Finally, they contain the organic compounds of the surface films, whose chemistry and interactions with atmospheric properties are still a vast area of study.

3.2.5. *Flux measurement*

To quantify surface fluxes, it is necessary to estimate their magnitude, directly (by measurements) or indirectly. The principal approaches used and contested issues are presented in this section.

3.2.5.1. *Radiative fluxes*

The measurement of radiative fluxes is made locally with specific instruments: the radiometer (for solar radiation), pyranometer (IR radiation) and pyrgeometer. These sensors are made up of materials that transform the input into an electric current inside a transparent cell. For example, in the pyrgeometer, the sensitive element that is exposed to infrared radiation is a thermopile, the solar radiation being eliminated by a non-reflective black paint. The thermopile delivers an electric potential difference proportionate to the difference between the received radiative flux and that emitted by the sensitive surface of the thermopile.

Radiation profiles in the ocean are measured by different sensors that can be operated from ships, or in automatic versions, installed on moored buoys or profiling floats: sensors for solar radiation or multifrequency radiative measurements, chlorophyll-A fluorescence or dissolved colored matter, as well as absorption profiles (by transmissiometry).

To estimate the surface radiative balance on a global scale, satellite-borne sensors have been developed. Some of them measure solar and infrared radiation directly at their flight altitude, enabling the planet's energy balance to be obtained. However, physical or statistical modeling of the relationship between this measurement and the surface balance is required to establish the radiative transfer throughout the atmosphere.

From satellites, the surface temperature of the oceans is obtained with sensors measuring the Earth's radiation in thermal infrared (but in a narrow band, contrary to the long wave radiation). Hypotheses for emissivity are formulated to obtain the surface temperature; the atmospheric absorption is removed using two similar bands of long waves (10 and 12 μm). However, the infrared radiation emission layer is extremely thin, less than one mm thick, and the temperature thus obtained differs from the temperature measured by the *in situ* sensors (a measurement performed at several tens of centimeters, down to more than a meter). The two quantities are compared using heat transfer models of the oceanic surface layer.

From space, it is possible to estimate the chlorophyll concentration in the ocean surface layer, using measurements from visible imagers in the blue-green spectrum: "ocean-color" instruments such as Sea-viewing Wide Field-of-view Sensors (SEAWIFS), or multiobjectives ones, such as Moderate Resolution Imaging Spectroradiometer (MODIS) and Medium resolution Imaging Spectrometer (MERIS). The major problem is the aerosol content in the atmosphere, which contaminates the surface measurement and influences the reading. Another difficulty, especially in coastal regions, is distinguishing the signal from the phytoplankton from the one of mineral particles or dissolved colored matter.

3.2.5.2. *Turbulent fluxes*

Direct measurements using the correlation method or turbulent dissipation method are only possible on ships or large moored buoys, because of the electric power, mass sensors and volume, and the high-data rate and storage required. The quality of the measurements is therefore affected by several factors: the physical obstacle due to the ship superstructure (the flow wake), and the movements caused by rolling, pitching and displacement. Moreover, few temperature and humidity sensors allow such measurement at sea, without being damaged by the effects of spray in a marine environment... Aerodynamic mass methods are, therefore, mainly used for local measurements. On a ship, the problem of the flow disturbance remains. In practice, the sensors should be placed on a mast as far as possible from the ship superstructures at the bow, and measurements can only be used for winds blowing toward the ship (at ±30°C). Present strategies point toward measurements on buoys or even on drones sailing on the water surface.

From space, the methods enabling turbulent fluxes to be estimated are based only on the use of simplified aerodynamic formulae (a neutral case, due to the lack of information on the atmospheric stratification). Scatterometers (radars pointing obliquely from one side or either side of the satellite's tracks), altimeters (along the track on the vertical) and microwave imaging radiometers (at frequencies of 10–18 GHz) enable the surface wind to be estimated; the scatterometer is the only instrument that also establishes wind direction. The general principle of this measurement is that the small-scale roughness (centimeter scale) caused by the wind friction on the sea modifies the backscattering (radars) and the emission (radiometers) of electromagnetic waves. The relationship between this roughness and the wind (at 10 m) is calculated taking a wind profile for neutral stratification. In practice, statistical relationships are established between local measurements and satellites, or with simulated data sets.

Concerning heat fluxes, the problem is even more complex: there are currently no means for estimating the atmospheric temperature near the ocean surface using satellites, and the methods for obtaining the low-level humidity are devalued by significant errors. The sensible heat flux, generally small over the sea (most often <50 Wm^{-2}), cannot be estimated from space. As for the latent heat flux (which often reaches 150–200 Wm^{-2}), the uncertainty is equal or more than 30% when satellite data for the wind, surface temperature (to obtain the surface humidity) and atmospheric humidity are combined...

To obtain the water balance (*E-P*), both the evaporation flux *E* and precipitation rate *P* must be measured. A new obstacle emerges: measuring precipitations over the oceans. On a buoy, sea spray makes the rain gauges unusable; on a ship, the disturbance in the flow renders this measurement very uncertain, unless a rain radar can be used. In any case, the spatio-temporal variability of precipitations is such that local measurements are not reliable for use on a regional scale. Estimation by satellite is once again the only reasonable alternative. However, a sensor (radar or radiometer) only receives radiation that has been backscattered or emitted by droplets being in the atmosphere at the overpass time. Indeed, the precipitations are mostly used as hourly or daily rates. The development of statistical methods, established with data that are most often based on simulations (rain rate and satellite measurements) or reconstructed using other instruments, implies a reliable relationship and a thorough validation, which are unfortunately impossible to realize over ocean. The eventual uncertainty is broadly from 50 to 100% on the precipitation rate.

The calculation of the CO_2 flux, the only gaseous component for which researchers have attempted to get a global estimation of the flux at the surface, relies on the measurement of the wind and the CO_2 partial pressure in the surface water. It is determined on ships by carrying out a spectrometric measurement of the carbon dioxide in the air, previously equilibrated with the seawater. On ships, other methods are being developed using automatic sampling, also enabling the quantity of inorganic carbon dissolved in the water to be determined (for example, as done by the observation service CARbone AUStral (CARAUS)). Approaches using indirect automatic methods for determining pCO_2 have also been developed in the 1990s for installation on buoys (CARbon Interface OCean Atmosphere (CARIOCA) system, by colorimetry).

The flux estimation is made by combining the *in situ* data with those of the wind and surface temperatures, as estimated from space, and, more recently, of the surface salinity to better estimate the exchange coefficient. The uncertainty on this estimation remains, however, in the order of 50–100%.

3.3. Examples of energy exchanges

3.3.1. *Tropical cyclones*

Tropical cyclones are among the sea's most intense meteorological phenomena. They develop over warm waters (temperatures higher than 26°C), although not at the Equator, since the Coriolis force is null there: it is this force that puts into rotation the convective systems from which cyclones originate.

Thus, they form on either side of the Equator, at around 10° latitude, and when mature are made up of bands of intense convective systems organized in a spiral around a calm center (the eye) (see Figure 3.4). Their spatial characteristics can change radically in just one or two days. Some cyclones move rapidly, whereas others remain fairly stationary. When they approach a continent, their trajectory can branch off abruptly toward higher latitudes, transforming them into intense storms in temperate zones, as they collide with low pressure systems circulating from West to East (fairly frequent in the North Atlantic).

The passage of a cyclone causes a surface water cooling (through vertical mixing above all, but also heat loss by evaporation), which acts, in turn, on the cyclone itself, and possibly contributes to its evolution. On average, for identical conditions, it appears that a cyclone that causes less cooling will grow more dramatically and spread less quickly than a cyclone that causes a strong cooling.

The impact of a cyclone on the ocean surface temperature is very variable: the regions where the mixed layers are not very deep will be more rapidly cooled, just as in cases where the thermal stratification beneath is strong. Depending on the duration of the cyclone's passage and its intensity, surface cooling variations reaching as far as factor 8 can be induced. The air–sea fluxes of momentum, heat and water vapor tend to be reinforced on the right side of the cyclone track (in the northern hemisphere), because of the addition of its rotation speed with its displacement speed. In the same place, the strongest sea currents are generated, the swells moving in the direction of the cyclone being greatly increased with the associated momentum flux.

The winds and air–sea fluxes inside cyclones remain largely unexplained. Often, *in situ* instruments do not function properly in winds greater than 40 ms^{-1}, and satellite scatterometer data are not fully exploitable in these structures. It has been possible to document the wind structures inside cyclones using airborne flights, some buoys, and also by using new instrumentations, such as the "Aeroclipper" drifting balloons from *Centre National d'Études Spatiales* (CNES) in 2007. More recently, radiometric data in L band (ESA's satellite mission Soil Moisture and Ocean Salinity (SMOS)) have also supplied interesting information on the wind distribution within cyclones, illustrating strong variations over less than 50 km in some structures.

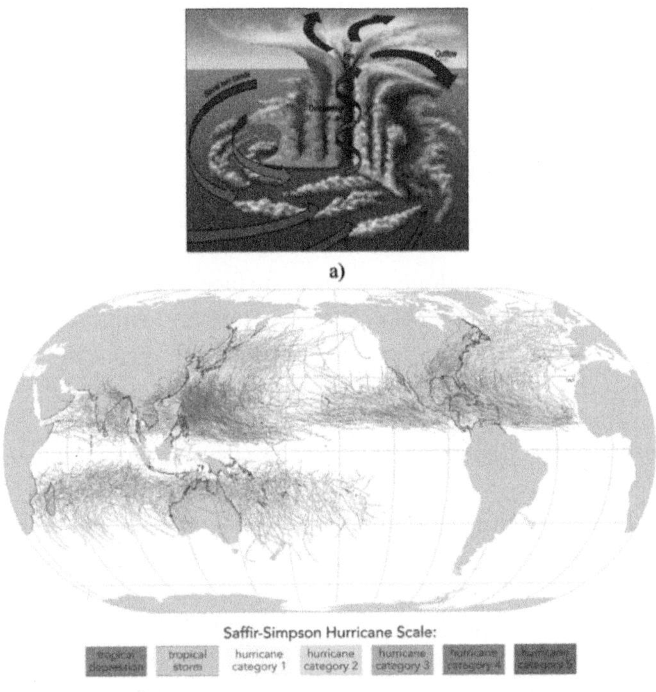

Figure 3.4. *a) The structure of a cyclone: the spiraled arm of the cumulo-nimbus surrounding the calm eye (taken from www.pitt.edu/~super7/20011-21001/20671.pdf) b) Trajectories of cyclones over the period 1945–2006. The color scale indicates the systems' intensity (http://commons.wikimedia.org/wiki/File:Tropical_cyclones_1945_2006_wikicolor.png, data from the Joint Typhoon Warning Center and the US National Oceanographic and Atmospheric Administration) (see color section)*

In such actively unstable conditions, with very strong currents and strong winds, the air–sea fluxes are not well known: getting near surface measurements is difficult, and the surface itself is not well-defined, because of the widespread presence of sea spray in the air and the penetration of air bubbles in the water. It is, therefore, mainly *a posteriori,* by comparing different numerical simulations of the sea surface temperature evolution, and through the means of heat balances, that it has been possible to precisely estimate fluxes and air–sea exchanges.

Moreover, the impact that an oceanic haline stratification can have on cyclones should be noted. This stratification is present at least seasonally in numerous regions where cyclones intensify. This is the case, for example, in the region East of the Antilles Islands, in summer and autumn, due to inputs of fresh water from the large South-American rivers; it is the case, too, in the Bay of Bengal, especially after the monsoon, when very large supplies of fresh water from the rivers penetrate far from the coasts into the open ocean. In these regions, haline stratification is often distinct from the thermal stratification. Here, the pull into deeper layers and the turbulent mixing at the base of the surface layer (whose thickness is restricted by the salinity) do not cause any surface cooling. In this case, the friction of the wind destroys the haline stratification but not the thermal stratification. This was observed using the SMOS mission, on several cyclones crossing the fresh water area in East of the Antilles Islands in 2010 and 2011 (at around 50 and 60 W) [GRO 12].

Cyclones are also associated with heavy precipitation. The Tropical Rainfall Microwave Mission (TRMM) space mission has permitted an exceptional documentation of the structure of many cyclones due to its precipitation radar (PR). The above-mentioned regions with strong evaporation are not in general in the same areas as the regions with heavy precipitation. Cyclones can thus be considered as redistributors of fresh water, and contribute locally to a haline stratification of the surface layers.

3.3.2. *Surface temperature fronts*

Regions of surface temperature (SST) horizontal gradients are called oceanic fronts. They are associated with a current or an eddy, and often reinforced by the sinking of surface water or the rising of water from the subsurface on one side of the front. These structures are, therefore, very common all around the globe, but with a high SST gradient variability, ranging from a few tenths of a degree to more than 10°C per 100 km for intense fronts.

For a long time, these phenomena have been neglected. The reason is that they remained unknown until the launch of space-borne fine resolution Infrared (IR) instruments around the end of the 1970s (see Figure 3.5). Moreover, global atmospheric and oceanic models have only recently been capable of representing such small–medium-scale phenomena, which were previously smoothed.

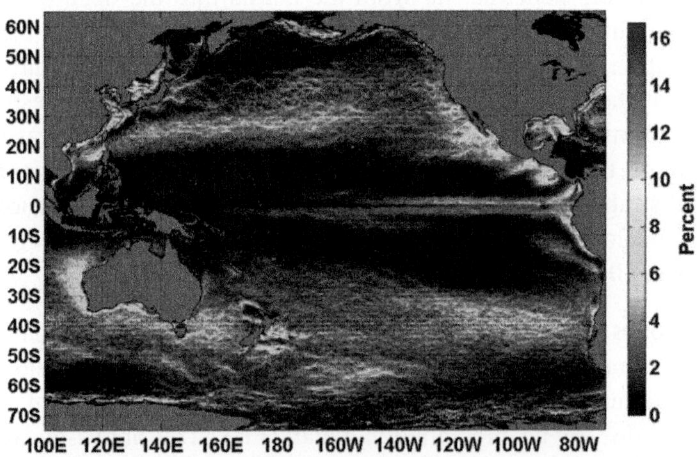

Figure 3.5. *Example of the detection of SST fronts using satellite images in infrared: an algorithm determines the probability for these fronts to occur, on average over the three months of winter (from [BEL 03]). The [CAY 92] algorithm is used (see color section)*

COMMENTARY ON FIGURE 3.5.– There is a strong likelihood of fronts occurring in the vicinity of large currents, and at the Equator, where they are associated with equatorial instability waves. SST fronts also appear in wide regions of the North and South Pacific basins, in the Southern Ocean and South Indian Ocean, exhibiting their "universal" distribution in the oceans.

However, studies carried out in 1980–1990 have shown that the atmospheric flow is modified while passing above an SST front, even if it has a weak intensity. In the case of a flow circulating from the warm to the cold side, the boundary layer stabilizes (heat flux decrease) and the wind slows due to the decrease in the layer thickness. Conversely, if the wind blows from the cold to the warm side, heat fluxes abruptly increase, leading to a deepening of the boundary layer, associated with an increase of the average wind. On a larger scale, a correlated decrease in surface pressure can be observed [SMA 08]. Indeed, the oceanic structures have a long-life duration (several weeks in the case of an eddy, months for currents (some are permanent)). The influence of these fronts on the atmosphere, when cumulated over the globe, is thought to have a significant impact: for example, Bourras *et al.* [BOU 04] have shown that the local wind beside an ocean vortex is disturbed as long as the ocean vortex remains. The SST fronts are, therefore, detectable not only through the difference in surface temperature, but also by the changes in surface roughness associated with the wind, by altimetry (dynamic changes associated with the sinking/rising of masses of water), and by changes in chlorophyll levels at the surface. A complete review of the detection of fronts and the oceanic mechanisms associated with them can be found in [ROB 10].

Coupled phenomena between ocean and atmosphere implying an SST front can be observed in all regions of the globe. The following shows the importance of strong SST fronts.

However, even fronts of moderate-to-weak intensity (less than 3–4°C) can play a role in the ocean–atmosphere coupling. For example, in the Gulf of Guinea, in the boreal spring, the trade winds of the Southern Hemisphere are pulsed by the Santa Helena high. These wind bursts trigger a vertical mixing in the ocean, leading to the formation of a seasonal cold tongue of fresh water. To the north of this strip, along the Equator, a discontinuity appears, with the sinking of surface water and the formation of an SST front. This front can produce changes of 2–3°C during spring.

Changes in the boundary layer and fluxes due to the wind blowing from the cool water to the warm water in the north of the gulf are associated with the development of this frontal structure. Thus, the ABL becomes thicker on the Northern than on the Southern side, and the flux and pressure changes lead to the development of a low-pressure area, favoring deep convection over the warm water in the North of the Gulf and in the vicinity of the coast. The preservation and stability of this phenomenon over several weeks is supported by an overturning atmospheric circulation in the troposphere, from North to South. Thus, convective precipitation is concentrated on the Guinean coast of Africa in spring, before shifting over the continent as far as Sahel [LED 12].

3.3.3. *The interactions close to strong oceanic fronts*

At 25°N, ocean and atmosphere contribute around the same proportion of heat transported toward the Poles. Seventy-percent of the oceanic contribution to this transport is lost between 25°N and 45°N, mainly in the regions of western boundary currents and during the season from December to March.

These regions are the location of great contrasts in ocean temperature, often with more than 10°C variation over 100 km across the ocean fronts that, on their side nearer the coasts, border the large seams of currents rising from the tropics that are the Gulf Stream and its branches and the Kuroshio/Oyashio, as well as the numerous ocean vortexs that detach from it. Strong fronts and strong ocean–atmosphere interactions are also found in the Southern Hemisphere, in certain sectors of the circumpolar current when it meets the Western boundary currents (such as the retroflection of the current of the Agulhas).

Later, we will focus on the interactions in the currents of the Northern Hemisphere, where three important campaigns have taken place in the course of the last years (Kuroshio Extension Study System (KESS), CLImate VARiability programme (CLIVAR) and Mode Water Dynamic Experiment (CLIMODE)). In the case of the Gulf Stream, there is also the contribution of the deep returning circulation from subpolar regions, and of nonlinear variability of these very intense western boundary currents. The displacement of the fronts causes major anomalies (at a fixed point) in temperature. In the first place and when the climate scales are integrated, the air–sea fluxes of heat associated lead to a negative local retroaction on the anomalies in temperature, a retroaction that tends elsewhere to damp the warm vortices and to isolate cold structures from the surface, such as the famous "rings" of the Gulf Stream. These fluxes, as well as the deeper thermoclines in some of these structures, contribute to deep OMLs in warm sectors.

During the seasonal restratification in spring, the waters of the deep mixed layers, which are called "mode" waters, are isolated from the surface and flow afterward into the subsurface in the recirculation of currents West of the subtropical gyres. Thus, the estimations made during CLIMODE indicate that approximately 50% of the mode water formed is in the vicinity of the oceanic fronts of the Gulf Stream and warm seams of currents.

The strong fluxes of sensible and latent heat (and therefore humidity) have a very important effect on the atmosphere. Locally, they increase the baroclinicity, can cause particularly deep convection, and favor, by baroclinic instablility, the development of atmospheric depressions. This fairly local effect has been well documented in numerous winter situations since the pioneering work carried out during the Fronts and Atlantic Storm Track Experiment (FASTEX) [GIO 01]. For these reasons, and also because of the retroactions caused by the atmospheric ocean vortexes on the average circulation, it is thought that the storm tracks tend to be located close to the eastern coasts of the North-American and Asiatic continents up to the point where they separate from the western boundary currents, at least on longer time scales (seasonal or longer).

Modeling studies (see Figure 3.6), for which the model resolution permits a coherent representation for the mechanisms associated with the front, have shown the persistence of intense localized precipitation over the Gulf Stream zones [MIN 08]. Note that this sometimes implies deeper convection in the atmosphere, in any case during the seasons when the surface temperatures are warmer.

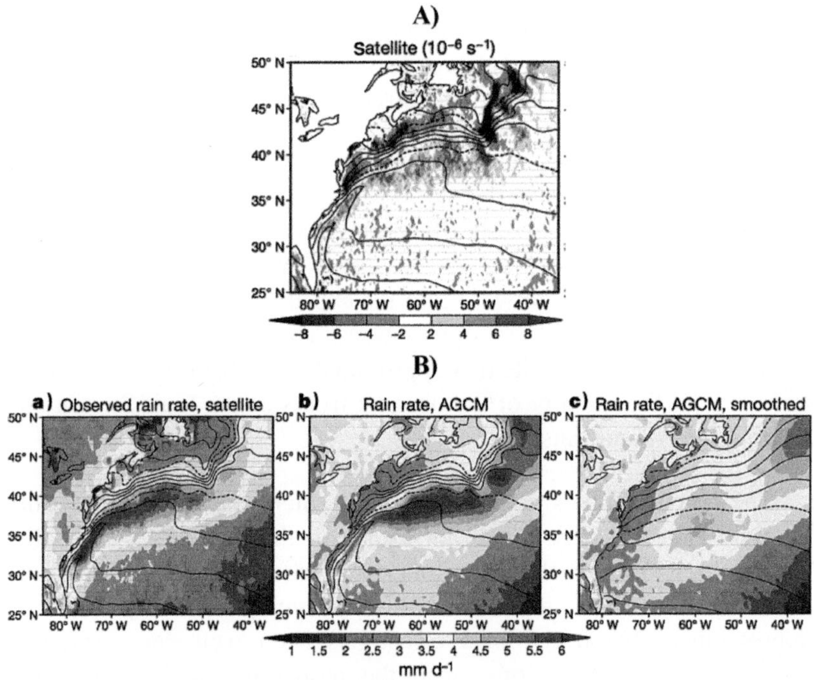

Figure 3.6. *Impact of the Gulf Stream on the atmosphere. A) Climatology of surface wind convergence and the surface temperature (contours), obtained by satellite B) Impact on precipitation: a) satellite observation, b) and c) precipitation simulated using a general atmospheric circulation model with a good enough resolution for describing the SST gradients (middle) or with a smoothed SST (right). From [MIN 08] (see color section)*

Thus, the western boundary regions, characterized by strong contrasts in oceanic temperature (just like the strong fronts along the circumpolar currents, in particular to the south-east of South Africa), are also regions of strong atmospheric variability during the winter season. Therefore, it is often possible to observe intense winds associated with cold, dry air coming from the continents. What are the exchange mechanisms that occur in these conditions? It has long been known, from the analysis of scatterometer winds in particular, that the surface winds and ABL properties are very different on either side of the strong fronts that border these currents or the ocean vortices that they generate.

These mechanisms are caused mainly by the contrasts in surface temperature, and therefore by the changes in the ABL. A high instability is observed over warm waters when continental air arrives. The strong winds and strong vertical gradients in temperature and humidity, therefore, induce very strong sensible and latent heat fluxes over the warm sectors, and it is not unusual for these combined heat fluxes to exceed 1,000 Wm^{-2}. As strong momentum fluxes are associated with sea waves, and sometimes with strong currents beside these fronts, this can lead to very heavy seas and to the formation of rogue waves. The intense winds and the resulting coolings reinforce the exchanges of gas, in particular of CO_2. The reason is that the vertical mixing caused by the strong momentum flux pulls waters rich in inorganic carbon into the surface layer, while the cooling would tend to diminish the partial pressure of CO_2 in the seawater. In the atmospheric boundary layer, the marked instabilities trigger a rapid deepening of this layer and eventually the formation of shallow convection. Associated clouds are, likewise, also very different on either side of the fronts.

3.3.4. *Marginal ice zones and associated air–sea fluxes*

Sea ice occupies the main part of the icy Arctic Ocean in winter, while its surface is reduced in summer. For the last 10 years, the amount of ice which is melting in summer has been significantly increasing (see Figure 3.7). The mechanism responsible for this reduction has not yet been established, since the number of *in situ* observations in the ocean, in the atmosphere and at the surface is clearly insufficient, and climatic models cannot satisfactorily reproduce the processes by which the ice forms and disappears.

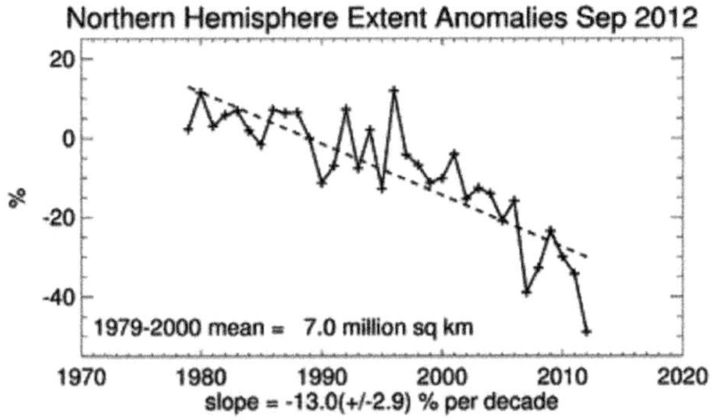

Figure 3.7. *Horizontal extension of the arctic ice field on average over the month of September (annual minimum anomaly in percentage) from 1979 to 2012 (National Snow and Ice Data Center)*

However, the feedback mechanisms involving sea ice are well known:

– the presence of sea ice evidently inhibits air–sea fluxes. After it has formed during winter, the lower atmosphere cools greatly through radiative loss. This cooling is reduced occasionally, by the arrival of warm air within low pressure systems. The ice remains for a long time, since its very high albedo prevents the solar energy from penetrating into its depth and the water beneath. But, particle deposition (in particular of black carbon transported from Asia) tends to reduce this albedo. Indeed, the more the surface albedo diminishes, the more the penetration of solar radiation increases, and warms the upper layers of the ocean, inhibiting the formation of ice even more (positive feedback). The ice formation at the beginning of the polar winter, and its persistence late into spring are, therefore, necessary for retaining an overall stable sea ice coverage from year to year. The contrary has been observed over the last 10 years;

– in marginal ice regions, in free water areas formed within the ice field under the effect of ice divergent movements, or also in channels opened by shear stress and ice floe deformation, flows of cold polar air induce very strong exchanges of sensible and latent heat. The lower atmosphere is very unstable, which can lead to the development of particularly thick atmospheric boundary layers if these free waters extend over 100 km or more: for example, a 2 km thickness has been measured in the Labrador Sea. The surface winds also tend to be reinforced, leading to more waves and swell, which act on the ice edges and contribute to their fracturing. When air flow comes from the Arctic Ocean, next to eastern Greenland, the strong air–sea fluxes are the cause of the density increase of the surface waters. Depending on the initial haline stratification, they could possibly pull in the waters beneath. In the Greenland Sea, deep convection movement can be initiated, favored by the partial ice formation and the brine ejected from this young ice. If the initial haline stratification is strong, due to the presence of fresh surface waters, this deep convection process is inhibited and the production of sea ice increases until an ice plate is formed. However, the warming of the Arctic Ocean due to the decrease of the ice coverage could lead to a reduction in the dense water formation, indeed according to some authors, to a decrease in the intensity of the North Atlantic Drift through a blockage in its northward circulation.

The Antarctic region is also undergoing the effects of the Earth's global warming, but with contrasting effects because of the presence of the Antarctic continent. The seasonal ice fields seem to have grown over the past few years, where the mechanism responsible for this is not well understood. In summer, they disappear almost entirely, but a melting of the coastal glaciers is also observed, which causes massive influxes of fresh water into the sea.

Observation of sea ice:

– the installation of measuring stations *in situ* is a technical challenge: outside of the perennial ice zones (at least those that survive over several years), the risks due to ice plates colliding are high. Moreover, because of the channel opening occurrence, floating instrumentation is required, that must be resistant to freezing sea spray. Finally, an additional, non-negligible risk is the material being destroyed by a polar bear;

– by satellite, visible observations (during the polar day) are often difficult because of the stratiform cloud cover. Using infrared sensors, measurement of the surface temperature can be wrong because of the slight difference between the temperatures emitted by the clouds and those from the surface. However, using microwaves, the radiometric observations have proved to be adequate for monitoring the ice "concentration", that is to say the quantity of the ice surface by unit of surface and its extension. The microwave emission is coming from an ice layer, whose thickness depends on its stratification and associated distribution of dielectric properties. This complexity has, hitherto, limited a more "physical" exploitation of the signal.

It is, therefore, a major challenge to monitor the Arctic and Antarctic sea ice, keeping in mind the importance of these regions for the Earth's climate. In 2007, the "International Polar Year" triggered an international observation effort, which is still being pursued.

3.3.5. *The example of the impact of the iron supply from Saharan aerosols*

The heart of the subtropical Atlantic Ocean is a region with an often deep thermocline which lacks any real ocean vortex activity that might cause important vertical exchanges. As a result, there are few supplies of nutritive salts coming from the deeper ocean or regions of *upwelling*, such as those off Africa. Therefore, except for a short period at the end of winter when the thermocline is close to the surface, weak concentrations of dissolved nutritive elements, in particular of nitrates and iron, are found at the surface in these regions. The supplies of aerosols, however, are significant and have different origins: anthropogenic aerosols rich in inorganic nitrated elements, above all in winter; natural aerosols coming from Saharan dusts; aerosols from the Sahel, especially from bush fires. These different aerosols do not reach the same oceanic regions.

The aerosols coming from Africa are very significant in volume (between 10 and 50 gm^{-2}/year from 5 to 25°N and east of 30°W), and are deposited with the trade winds as far as the Antilles and even the extreme southeast of the North-American continent. They are transported within the lower and middle atmosphere and reach the ocean through dry deposits or humid deposits, during convective rains. Their rate of spatio-temporal deposition is very variable, and it is estimated that in certain regions, a single event can contribute to almost half the natural annual supply. They certainly constitute an important source of soluble, bioavailable iron, which is usable fairly directly by plankton. They also contain a small amount of phosphorus (0.08% of their mass, about 1/3 is bioavailable), and relatively few nitrates (0.1% of their mass). They, therefore, have the potential to enrich the oceanic surface layers in nutrients enabling the development of certain groups of phytoplankton, in particular for tiny phytoplankton (picophytoplankton), which constitute an important part of the biomass in these regions. Their effect is more important on the development of diazotrophes, which use nitrogen dissolved in water, in place of nitrates, as a nutritive element, and will therefore be less constrained by the weak concentrations in nitrates from these aerosols. Experiments carried out during campaigns *in situ*, in particular in the summer season, with tropical water to which is added the Saharan aerosols, have shown an increase in the primary production of diazotrophic organisms, suggesting that these are colimited by the availability of iron and phosphorus of Saharan origin.

On a global scale, knowing the impact on the production of the system is more complex, since stimulation by Saharan aerosol supplies also affects the bacterial activity of the oligotrophic regions. Thus, the accumulation of organic carbon dissolved in the surface layers over the course of the summer season could be attributed to some of these responses where organic carbon would ordinarily be consumed by bacterial activity. The "biomass pull" effect of solid particles, when these cross the column of water, has also been mentioned. The question is to discover if the dissolving speed, total or partial, of the solid particles is more rapid than the secondary "biomass pull" effects, these probably being weak in the very oligotrophic regions of the oceans. Moreover, it is necessary not to overstate the importance of supplies of Saharan aerosols, knowing that currently the production in these regions is probably caused more by the nitrogen input from anthropic aerosols than by diazotrophic activity.

3.4. Conclusion

The ocean–atmosphere interactions are central in the climate functioning, through their role in the transfer of energy and matter between both systems. They are involved in small-scale processes but with spatial and temporal variations on all scales. These complex processes involve the physics, chemistry and biology inside extremely variable turbulent phenomena, bubbles and waves. Surface fluxes are also difficult to measure and quantify precisely because of the immensity of the oceanic regions. However, recent work (particularly, since the 1990s) has permitted the consolidation of simple parameterization approaches enabling fluxes to be estimated using average data, whether they are measured *in situ* or from space, while new, more precise instrumented platforms (moored buoys and surface floats, either drifting or remotely guided) are coming into use.

These developments essentially concern local processes. The transfer to larger scales (those that are represented in coupled ocean–atmosphere models) remains open to discussion: how are the exchange processes in space and time aggregated, taking into account different scales of variability in the ocean and atmosphere? What errors occur through the use of simplified local formulations in the study or modeling of the mechanisms of interseasonal to interannual climatic variability?

If satellite observations shed light on otherwise inaccessible information, the retrieved variables are likely to be obtained with insufficient precision, and are hence often of poor spatial and temporal resolution. Areas for improvement range from the sensors, to their use in models, but the question of how representative this information is for the ocean–atmosphere interface is crucial, in terms of precision and the scale of observation.

The challenges for the coming years are to be defined by the joint use of available information. The assimilation tools for the data and modeling will be essential in making the link between local and regional scales. Already, approaches which use coupled models on a fine scale for particular climatic regions are being developed to understand the causes of systematic biases in climatic simulations.

Determining the ocean–atmosphere interactions in some regions represents an important goal for climate understanding. Among them, coastal regions have a particular importance, especially in upwelling regions in the eastern ocean basins, where the scales, both oceanic and atmospheric, are the smallest, and which present a strong variability, while being critical regions in terms of air–sea exchanges. The polar oceanic regions and above all those that are partially covered with ice, in the Arctic and Antarctic, are also key regions due to their strong but still poorly understood climate evolution. New approaches will probably be required in this case, since the scale integration there is particularly complex due to the fully nonlinear processes.

3.5. Bibliography

[BEL 03] BELKIN I.M., CORNILLON P., "SST fronts of the Pacific coastal and marginal seas", *Pacific Oceanography*, vol. 1, pp. 90–100, 2003.

[BLA 87] BLANC T., "Accuracy of bulk-method-determined flux, stability, and sea surface roughness", *Journal of Geophysical Research: Oceans*, vol. 92, no. C4, pp. 3867–3876, 1987.

[BOU 04] BOURRAS D., REVERDIN G., CANIAUX G., *et al.*, "Response of the atmospheric boundary layer to a mesoscale oceanic eddy in the northeast Atlantic", *Journal of Geophysical Research*, vol. 109, 2004.

[CAY 92] CAYULA J.F., CORNILLON P., "Edge detection algorithm for SST images", *Journal of Atmostpheric and Oceanographic Technology*, vol. 9, pp. 67–80, 1992.

[DA 94] DA SILVA A.M., YOUNG C.C., LEVITUS S., "Atlas of surface marine data 1994, algorithms and procedures", *NOAA Atlas NESDIS 6*, Silver Spring, vol. 1, 1994.

[GIO 01] GIORDANI H., CANIAUX G., "Sensitivity of cyclogenesis to sea surface temperature in the Northwestern Atlantic", *Monthly Weather Review*, vol. 129, no. 6, pp. 1273–1295, 2001.

[GRO 12] GRODSKY S.A., REUL N., LAGERLOEF G., *et al.*, "Haline hurricane wake in the Amazon/Orinoco plume: AQUARIUS/SACD and SMOS observations", *Geophysical Research Letters*, vol. 39, no. 20, 2012.

[LE 03] LE QUÉRÉ C., *et al.*, "Two decades of ocean CO_2 sink and variability", *Tellus B*, vol. 55, no. 2, pp. 649–656, 2003.

[LED 12] LEDUC-LEBALLEUR M., DE COETLOGON G., EYMARD L., "Influence of the sea surface temperature on the coastal rainfall in the Gulf of Guinea during boreal spring", *Quaterly Journal of the Royal Meteorological Society*, 2012.

[LIS 86] LISS P.S., MERLIVAT L., "Introduction and synthesis", in BUAT-MÉNARD P. (ed.), *The Role of Air-Sea Exchange in Geochemical Cycling*, Springer, New York, pp. 113–127, 1986.

[MIN 08] MINOBE S., KUWANO-YOSHIDA A., KOMORI N., *et al.*, "Influence of the Gulf Stream on the troposphere", *Nature*, vol. 452, pp. 206–209, 2008.

[ROB 10] ROBINSON I., *Discovering the Ocean from Space: The Unique Applications of Satellite Oceanography*, Springer Praxis Books, Berlin, 2010.

[SMA 08] SMALL R.J., *et al.*, "Air-sea interactions over ocean fronts and eddies", *Dynamics of Atmosphere and Oceans*, vol. 45, pp. 274–319, 2008.

[WAN 92] WANNIKHOF R., "Relationship between gas exchange and wind speed over the ocean", *Journal of Geophysical Research*, vol. 97, pp. 7373–7381, 1992.

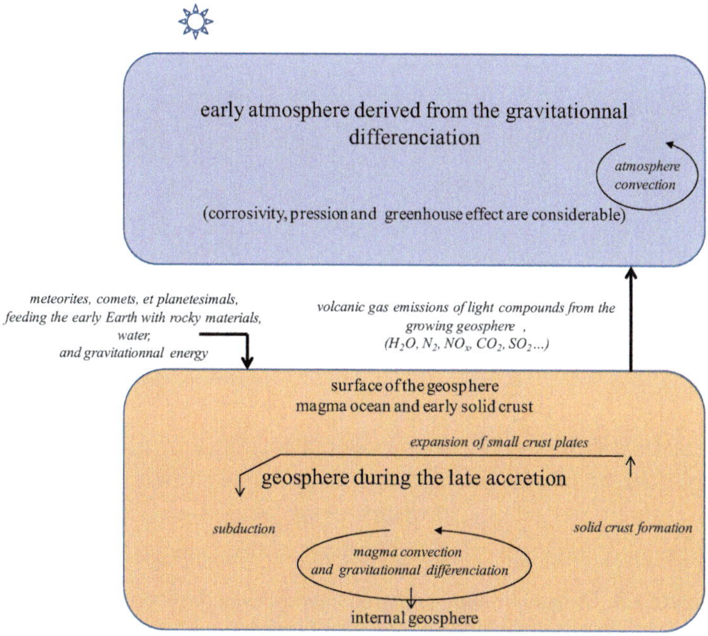

Figure 1.1. *Diagram of the Earth system at the end of accretion around 4.5 billion years ago*

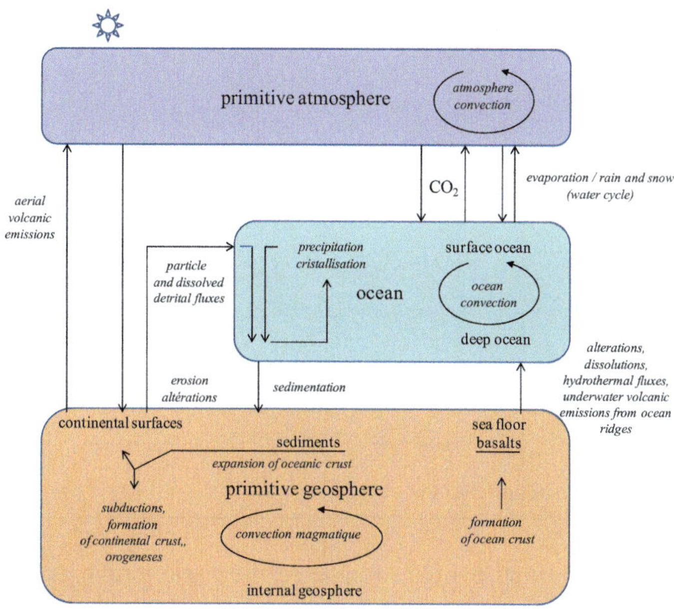

Figure 1.2. *Diagram illustrating the bases of the first abiotic regulations of the Earth system*

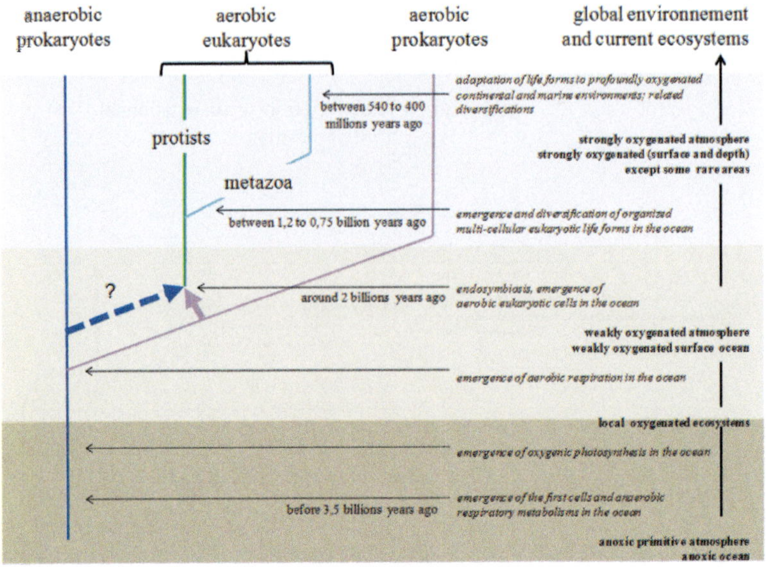

Figure 1.3. *The main evolutionary developments and the major environmental evolutions between the beginnings of life on Earth (over 3.5 billion years ago) and 400 million years ago (early Devonian Period)*

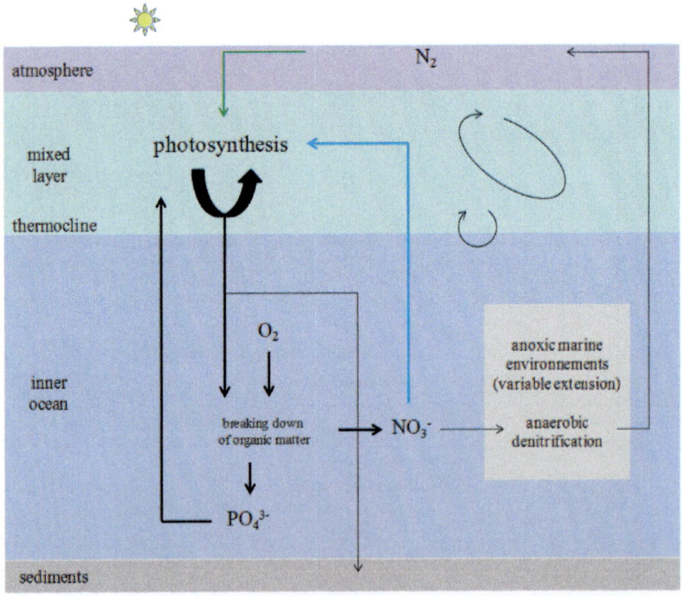

Figure 1.5. *The principle of the regulation of oceanic photosynthesis across the rapid cycle of major nutrients, nitrogen and phosphorus*

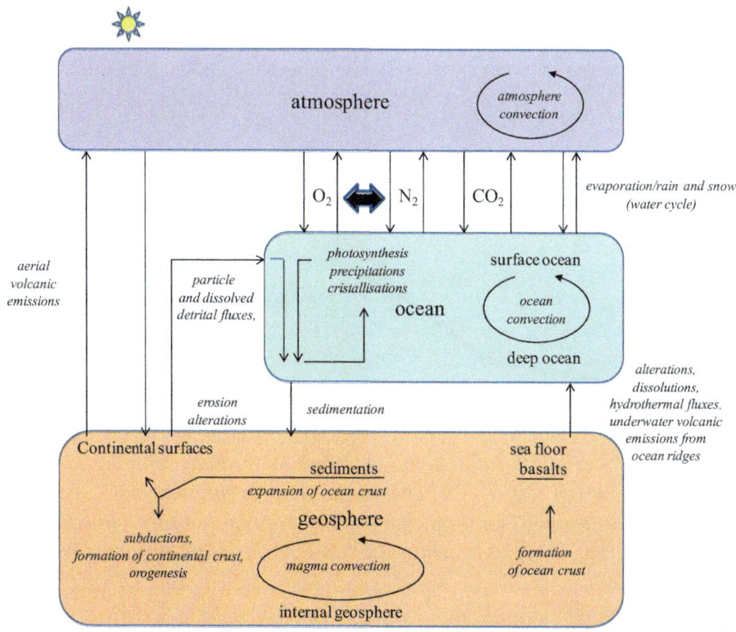

Figure 1.6. *The Earth system as it has functioned since the massive colonization of the ocean by photosynthetic life forms producing oxygen, that is to say for more than 2.5 billion years*

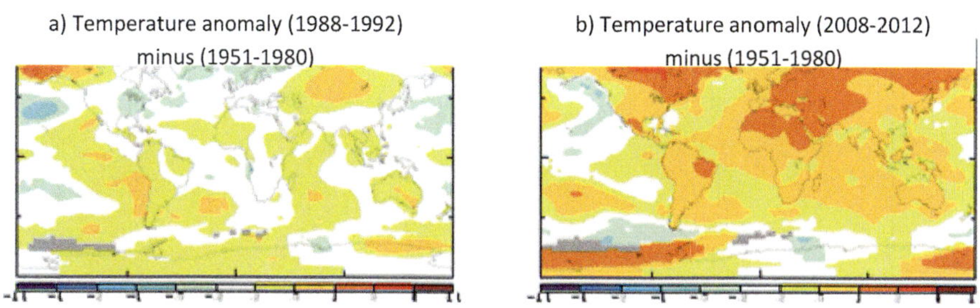

Figure 2.2. *Distribution of anomalies in temperature on the Earth's surface for 1988–1992 and 2008–2012 compared to the reference period 1951–1980 (according to http://data.giss.nasa.gov/gistemp/)*

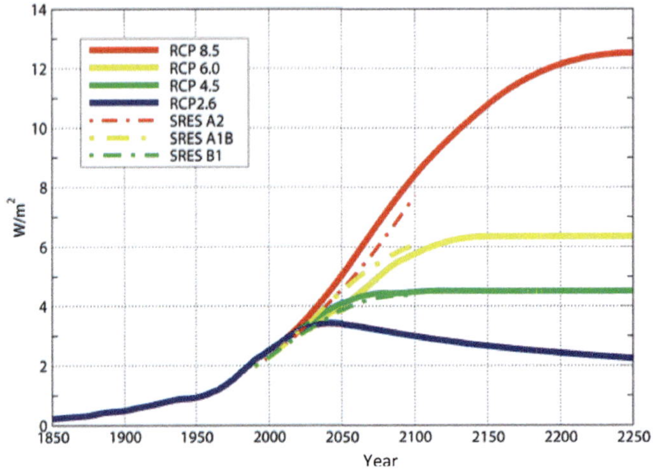

Figure 2.4. *Evolution of the Earth's radiative budget or "radiative forcing", in Wm^{-2} for the period 1850–2250. From 2006, the continuous lines represent new scenarios called Representative Concentration Pathways; the dotted lines represent the old scenarios (IPCC, 2007 and 2001) (source: ©Stéphane Senesi (CNRM-GAME, Météo-France/CNRS))*

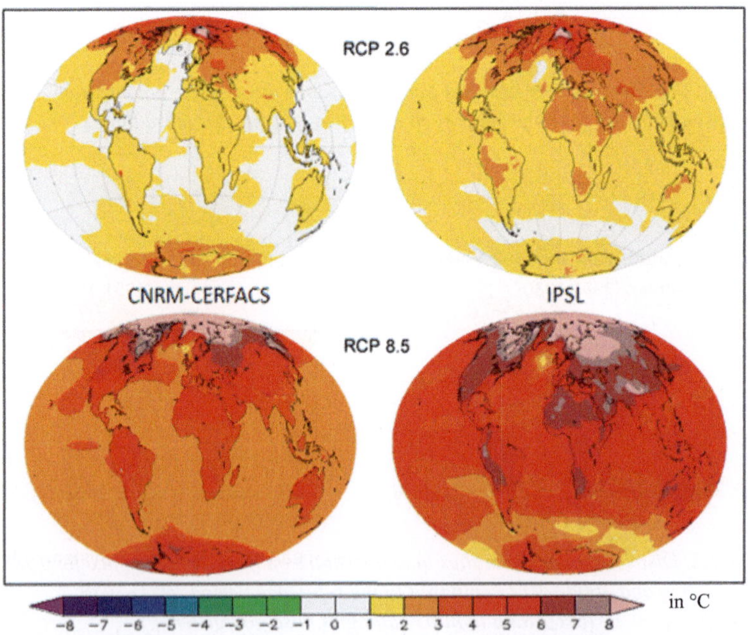

Figure 2.5. *Projection of the global temperature anomalies for 2071–2100 compared to 1961–1990 for model CNRM-CERFACS on the left and model IPSL on the right, for RCP 2.6 above and RCP 8.5 below (©Patrick Brockmann (LSCE/IPSL, CEA/CNRS/UVSQ))*

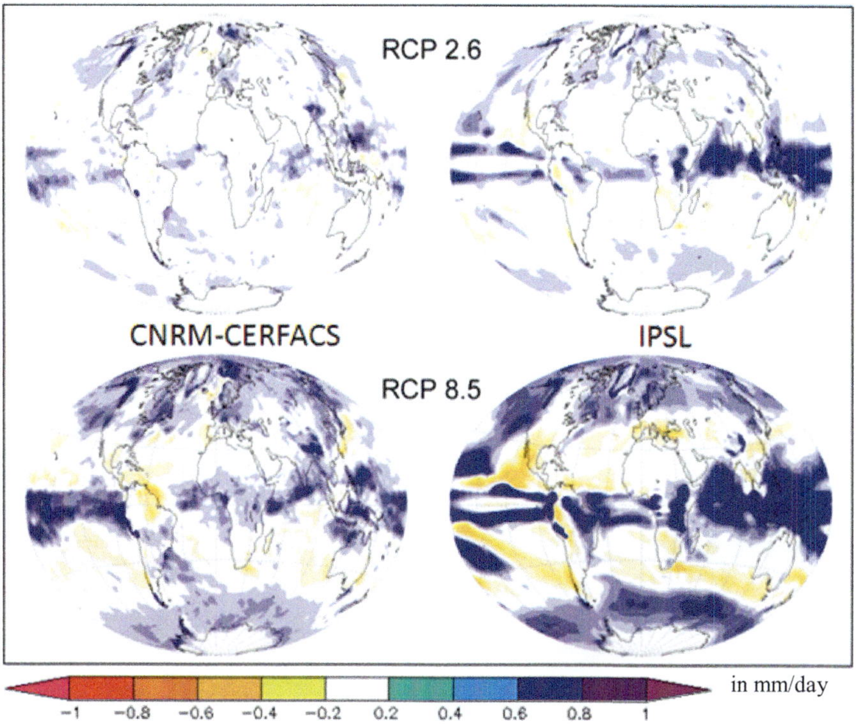

Figure 2.6. *Changes in precipitation in the period 2071–2100 compared to those in the period 1971–2000 calculated from models by CNRM-CERFACS and IPSL, for RCP2.6 above and RCP8.5 below (©Patrick Brockmann (LSCE/IPSL, CEA/CNRS/UVSQ))*

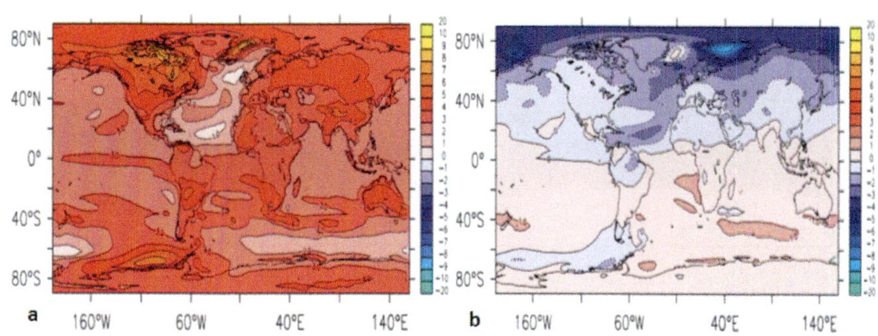

Figure 2.24. *Anomalies in annual global temperature calculated by the coupled ocean-atmosphere model in a simulation in which the concentration of CO_2 is doubled with an imposed melting rate for ice. Map of the difference after 500 years with the temperature of the model for the present state a), map of the difference after 500 years with a coupled simulation without ice melting b) (source: Didier Swingedouw (LSCE/IPSL, CEA/CNRS/ UVSQ))*

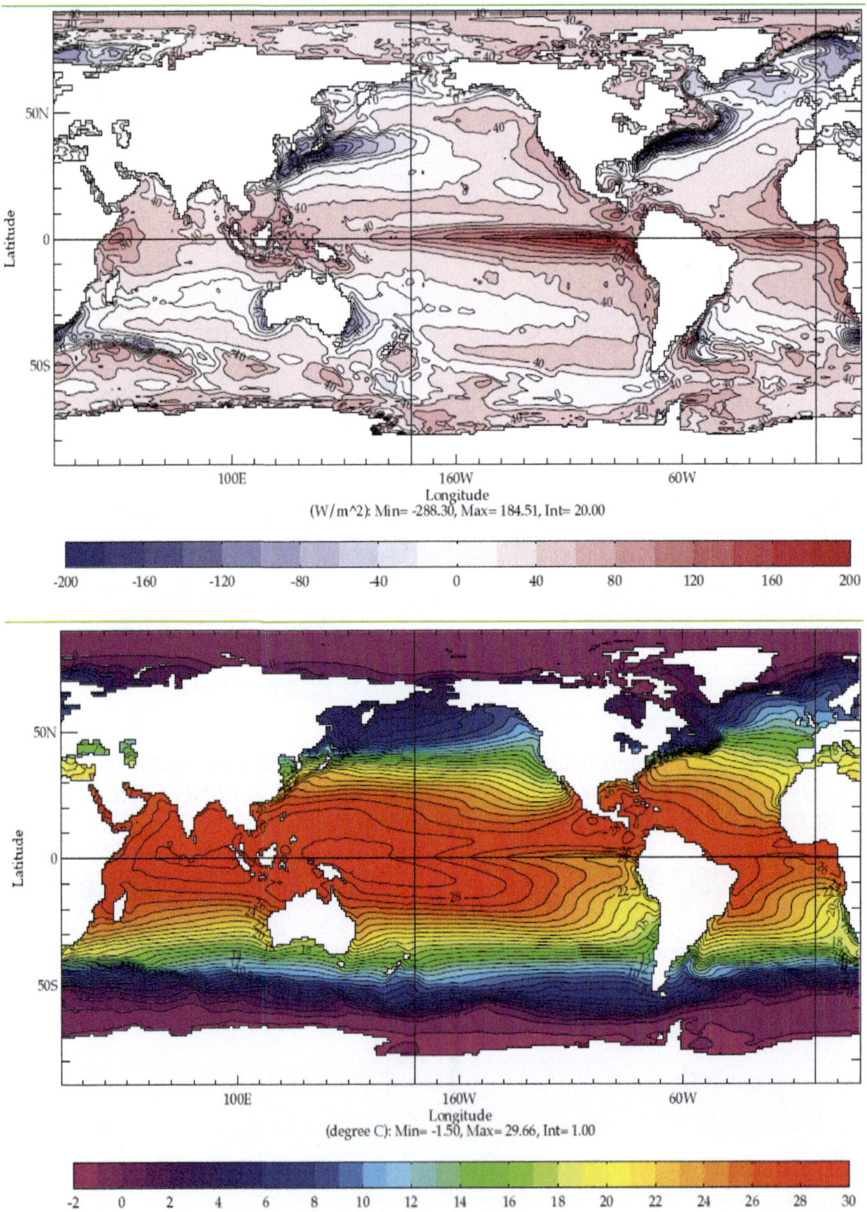

Figure 2.14. *Mean annual heat budget at the ocean–atmosphere interface from 1984 to 2007 in W/m² above and, below, mean annual temperature from 1984 to 2007 on the surface of the oceans in °C (after data from the site http://oaflux.whoi.edu/heatflux.html) © Sébastien Masson (LOCEAN/IPSL, CNRS/IRD/UPMC)*

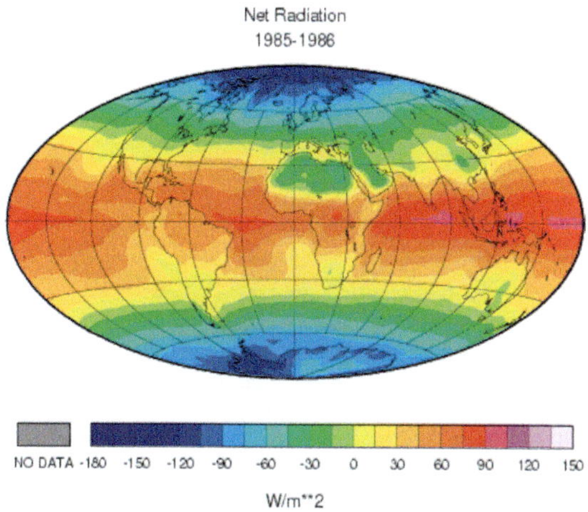

Figure 3.1. *Net annual mean radiative flux (data from the satellite ERBE, http://eos.atmos.washington.edu/cgi-bin/erbe/disp.pl?net.ann.)*

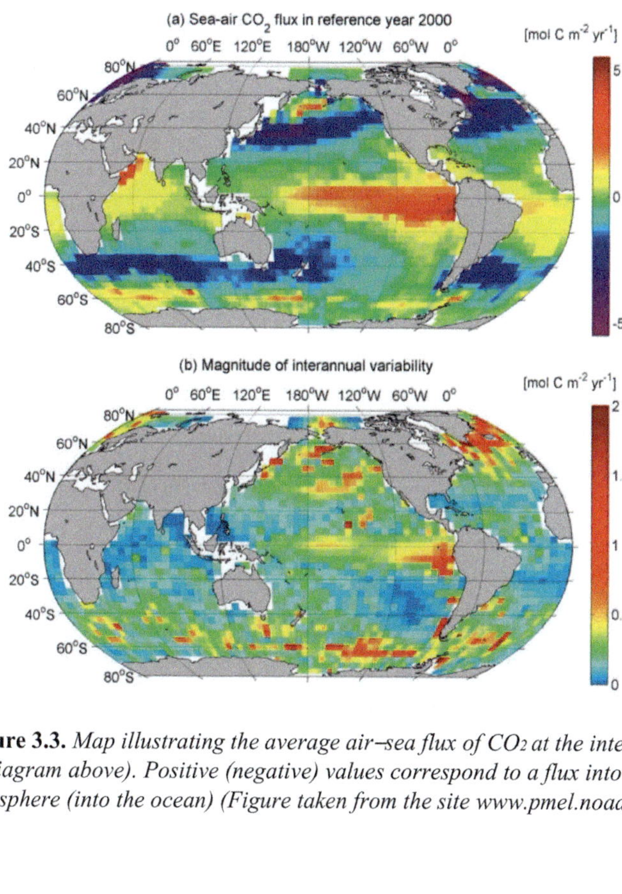

Figure 3.3. *Map illustrating the average air–sea flux of CO_2 at the interface (diagram above). Positive (negative) values correspond to a flux into the atmosphere (into the ocean) (Figure taken from the site www.pmel.noaa.gov/)*

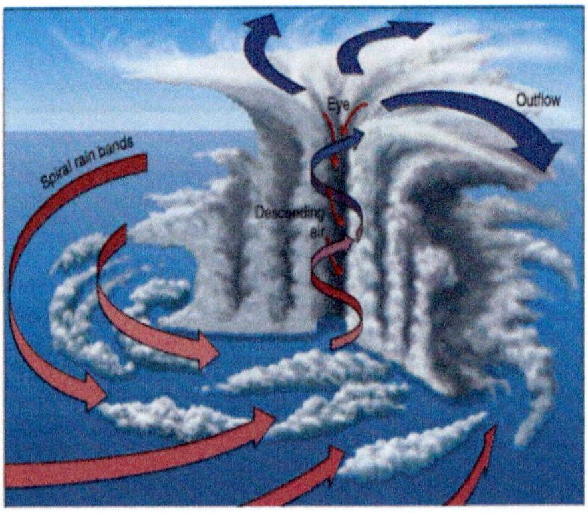

a)

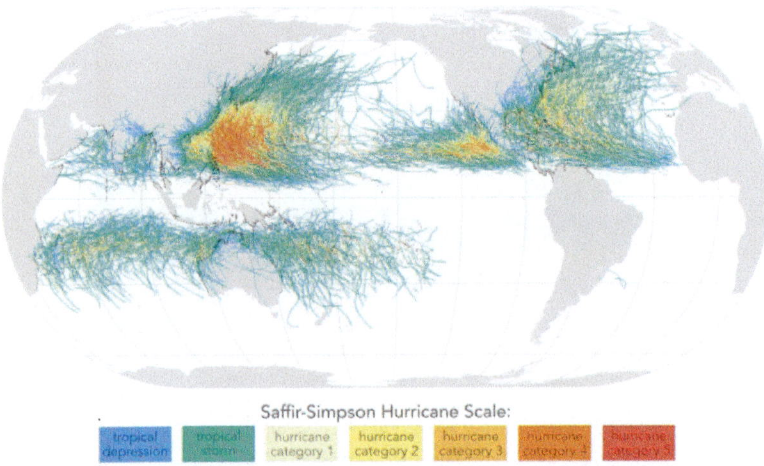

b)

Figure 3.4. *a) The structure of a cyclone: the spiraled arm of the cumulo-nimbus surrounding the calm eye (taken from www.pitt.edu/~super7/20011-21001/20671.pdf); b) Trajectories of cyclones over the period 1945–2006. The color scale indicates the systems' intensity (http://commons.wikimedia.org/wiki/File:Tropical_cyclones_1945_-2006_wikicolor.png, data from the Joint Typhoon Warning Center and the US National Oceanographic and Atmospheric Administration)*

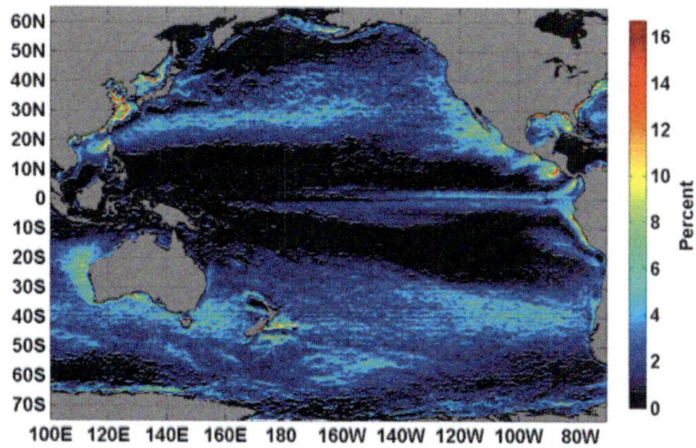

Figure 3.5. *Example of the detection of SST fronts using satellite images in infrared: an algorithm determines the probability for these fronts to occur, on average over the three months of winter (from [BEL 03]). The [CAY 92] algorithm is used*

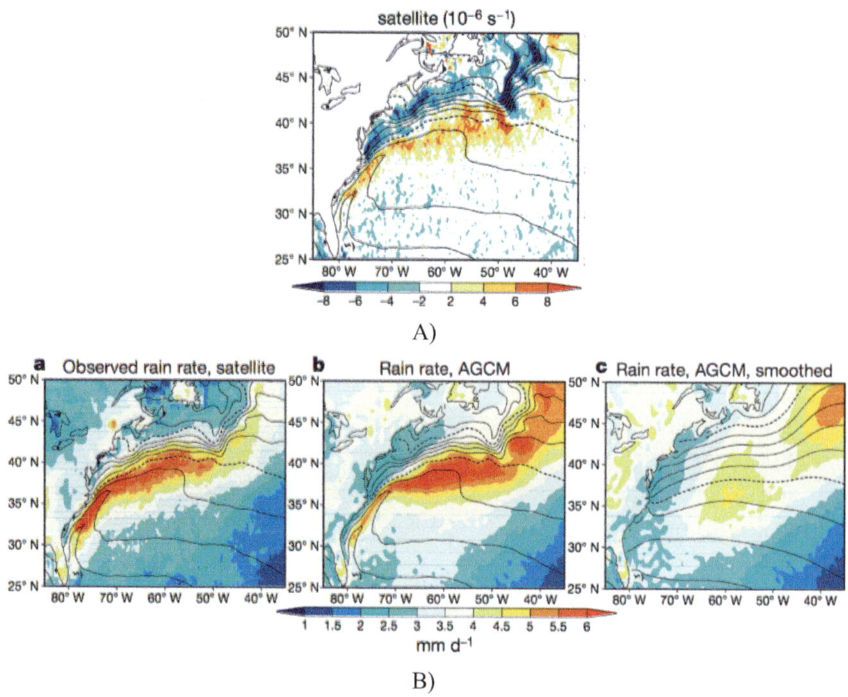

Figure 3.6. *Impact of the Gulf Stream on the atmosphere. a) Climatology of surface wind convergence and the surface temperature (contours), obtained by satellite b) Impact on precipitation: a) satellite observation, b) and c) precipitation simulated using a general atmospheric circulation model with a good enough resolution for describing the SST gradients (middle) or with a smoothed SST (right). From [MIN 08]*

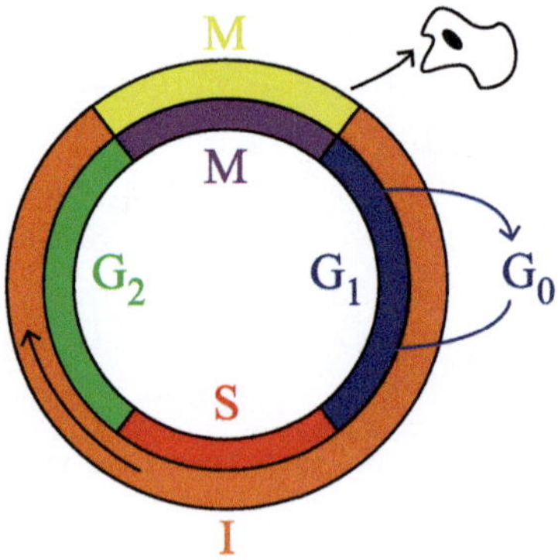

Figure 4.2. *Illustration of the cell cycle whose different phases are: G_0 (Gap 0, resting phase), G_1 (Gap 1, cell growth), S (synthesis, DNA replication), G_2 (Gap 2, cell growth), M (Mitosis, cell division) source: https://upload.wikimedia.org/wikipedia/commons/c/c4/Cell_cycle.png?uselang=fr, original figure downloaded from this site by Magnus Manske*

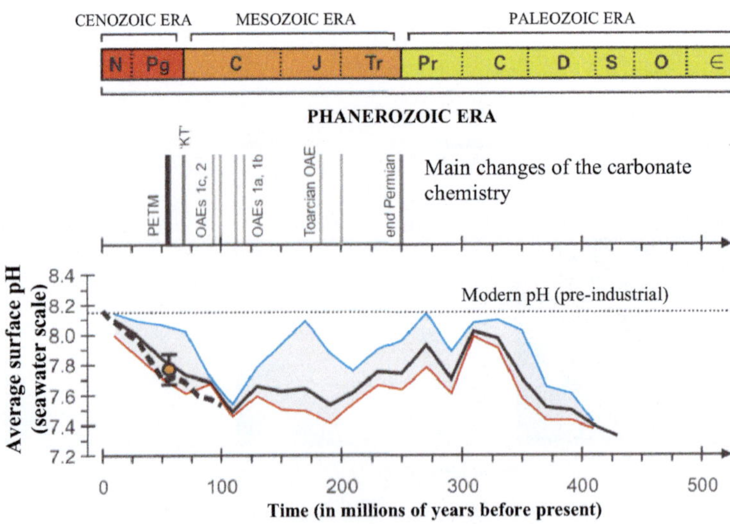

Figure 5.3. *Geological context of the acidification of oceans (according to [ZEE 11]): the major periods of acidification and the corresponding pH of surface waters on the "seawater scale"*

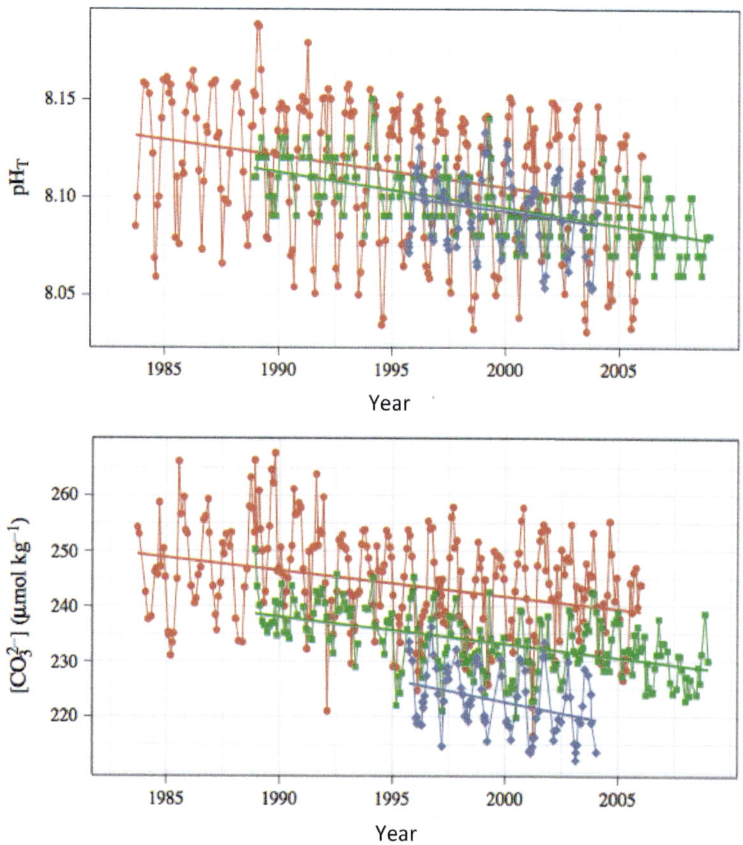

Figure 5.4. *pH (expressed on the total scale) and concentration in carbonate at the surface of the ALOHA (in green; [DOR 09]), BATS (in red; [BAT 07]) and ESTOC stations (in blue; [GON 10]) (according to [ORR 11])*

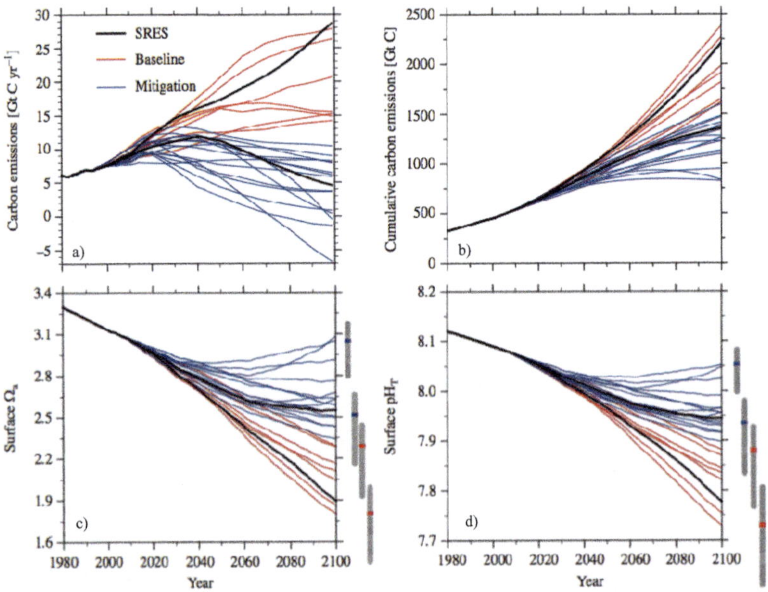

Figure 5.5. *Annual a) and cumulative b) CO_2 emissions used in the Bern2.5CC model and the resulting saturation in aragonite c) and pH_T d). The base scenarios are indicated in red, whereas the mitigation scenarios are shown in blue. The scenarios with high (SRES A2) and weak (SRES B1) CO_2 emissions are in black. The gray line shows the uncertainties in 2100 for the RCP scenarios. According to [JOO 11]*

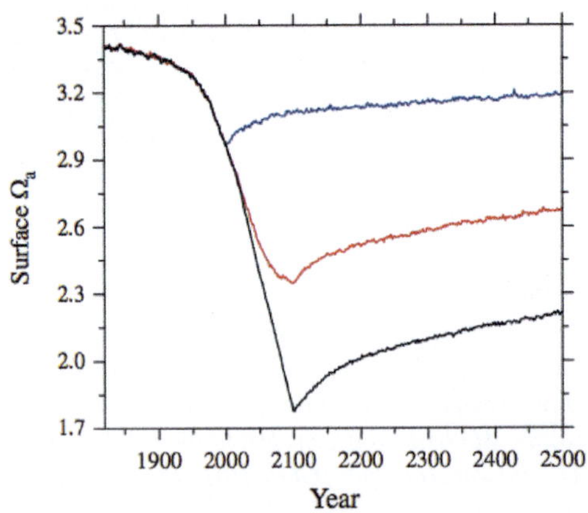

Figure 5.6. *Long-term impact of CO_2 emissions of the 21st Century on the saturation in aragonite estimated by the NCAR CSL1.4-carbon model. The three scenarios used are the following: no CO_2 emission from 2000 (blue), high emissions according to SRES A2 until 2100 and no emission afterward (red) and moderate emissions according to SRES B1 until 2100 and no emission afterward (black). According to [JOO 11]*

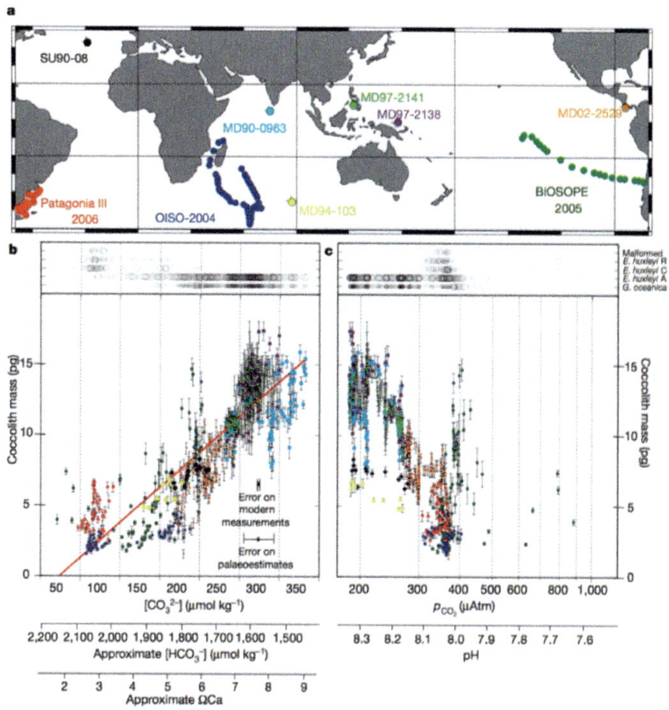

Figure 5.7. *Relationship between the mass of coccoliths from a large collection of samples covering different oceans: a) map of the locations of the samples and different timescales, and the carbonate chemistry; b) concentration in carbonate ions; c) partial pressure in CO_2 (source: [BEA 11])*

Figure 5.10. *a) Figure: photograph of a Posidonia bed next to a CO_2 vent (Castello Aragonese, Ischia, Italy; ©J. Hall-Spencer). b) Figure: top one shows changes in pH along a transect toward CO_2 vents; bottom figure shows along the same transect, percent cover (left axis) of calcareous (bold line) and non-calcerous algae (dots) as well as abundance (right axis) of sea urchins (dotted line). Data provided by Hall-Spencer et al. [HAL 08]*

4

Marine Biogeochemical Cycles

4.1. Introduction: geochemistry, biogeochemistry and marine biogeochemistry

4.1.1. *Geochemistry and cosmochemistry*

Geochemistry [JAM 09] forms part of the Earth Sciences. This discipline focuses on the chemical composition of the Earth and its constituents, in particular rocks and minerals, and aims to understand the cycles within which most chemical elements naturally present on our planet (there are over one hundred of them) and some of their components (e.g. water and carbon dioxide) circulate between the reservoirs that form the outer envelopes (or spheres) of the Earth. These reservoirs are the Earth's crust, the hydrosphere (which includes the oceans), the atmosphere and the biosphere[1].

We will not consider the interior of the Earth in this chapter, because its constituents (core and mantle) do not directly interact with the outer envelopes. Table 4.2 shows that the main chemical elements are not the same in the Earth's crust, the seawater and the atmosphere.

Geochemistry also focuses on the origins and evolution of chemical elements in the Universe, especially in the Solar System since this is where the Earth can be found. Here, geochemistry

---

Chapter written by Louis LEGENDRE.

1 Abbreviations and special symbols are summarized in Table 4.1.

becomes cosmochemistry. The main evidence of the chemical composition of extraterrestrial bodies is provided by meteorites that regularly fall to Earth. The only other physical samples of celestial bodies in our possession are those collected from the Moon by astronauts in the seven Apollo missions (USA) from 1969 to 1972, and those collected by automatic probes that have brought samples back to Earth, i.e. Luna 16 (Soviet Union) which landed on the Moon in 1970 and Stardust (USA) above the comet Wild 2 in 2004, which brought dust samples back to Earth in 2006. Other probes returning samples from other extraterrestrial bodies back to Earth are expected over the next few decades. In addition, there may be human expeditions to the Moon or Mars.

Abbreviation	Definition
Anammox	Anaerobic ammonium oxidation
DIC	Dissolved inorganic carbon
PIC	Particulate inorganic carbon
OMZ	Oxygen minimum zone
pCO_2	Partial pressure of CO_2
UV	Ultraviolet radiation
$\sum CO_2$	Total CO_2 or DIC (equation [4.2])

Table 4.1. *Abbreviations and special symbols used in this chapter*

Earth's crust		Seawater		Atmosphere	
Element	%	Element	%	Element	%
Oxygen (O)	46	Oxygen (O)	86	Nitrogen (N)	75
Silicon (Si)	28	Hydrogen (H)	11	Oxygen (O)	23
Aluminum (Al)	8.1	Chlorine (Cl)	1.9	Argon (Ar)	1.3
Iron (Fe)	5.0	Sodium (Na)	1.1	Carbon dioxide (CO_2)	<0.1

Table 4.2. *Names and percentages by mass of the four main chemical elements present in three major geochemical reservoirs that form the main outer envelopes of the Earth (i.e. the Earth's crust, the seawater and the atmosphere)*

COMMENTS ON TABLE 4.2.– The percentage given for each element is the fraction of the mass of that element in the total mass of the reservoir. For the atmosphere, each value takes into account the density of the gas and the air at normal temperature and pressure (20°C, 1 atm). In many geochemistry textbooks, the percentages cited for O and H in seawater are much lower than those given here because the authors did not take into account the large amounts of these two elements present in H_2O molecules.

Biogeochemistry is the area of geochemistry that focuses on the effects of living organisms on the concentrations of chemical elements in the different natural reservoirs of the Earth and on the cycles of chemical elements between reservoirs. During the last few years, the concept of biogeochemistry has widened and is now often referred to as “biogeosciences”, which focus on the interactions between the biological, chemical and physical processes in terrestrial or extraterrestrial life with the geosphere, hydrosphere and atmosphere (this definition is from the scientific journal *Biogeosciences*, published by the scientific society *European Geosciences Union*).

4.1.2. *Biogeochemistry and marine biogeochemistry*

The origin of the word “biogeochemistry” is interesting. An article on the history of biogeochemistry [GOR 91] reports that this word was used for the first time by the scientist Vernadsky (Figure 4.1) in his book *Biosfera*, published in Russian in 1926 [VER 26]. However, the author of the article had only seen an abbreviated English translation of this work, entitled *The Biosphere*, which was published in the United States in 1986 [VER 86]. Actually, Vernadsky translated his own Russian book into French in 1929, with the title *La Biosphère* [VER 29], which included both a translation (with additional contributions from the author) of the Russian book of 1926 and, in a second part, a translation of an oral communication he had presented in 1928 at the Society of Naturalists of Leningrad (now called St Petersburg). It must be noted that the terms “*biogéochimie*” (biogeochemistry) or “*biogéochimique*” (biogeochemical) do not appear in the French translation of the 1926 Russian book (pages 1–201 of the 1929 work), indicating that the author did not use these words in 1926, whereas the adjective “*biogéochimique*” is used several times in the translation of the oral communication of 1928 (pages 203–230 of the 1929 work). Hence the term “*biogéochimique*” was published for the first time in 1929, in the French language book of Vladimir Ivanovich Vernadsky [VER 29] and not in his Russian work of 1926 [VER 26] as often thought.

Marine biogeochemistry is the area of biogeochemistry that focuses on the effects of the activity of marine organisms on the concentrations of chemical elements in the biosphere, the hydrosphere and the geosphere and on the cycles of chemical elements between these reservoirs. Since the end of the Joint Global Ocean Flux Study (JGOFS) international program in 2003, the boundary between biological oceanography and marine biogeochemistry has become blurred and several areas of marine research, such as the photosynthetic production of organic matter from inorganic compounds by phytoplankton, known as primary production, are covered by the two disciplines. Key aspects of marine biogeochemistry are discussed in the book *Ocean Biogeochemical Dynamics* [SAR 06].

Figure 4.1. *Vladimir Ivanovich Vernadsky, 1863–1945, founder of biogeochemistry (https://en.wikipedia.org/wiki/File:Vernadsky.jpg)*

The dissolution of the main atmospheric gases into seawater provides an example of the differences that exist between the ocean and other biogeochemical reservoirs. In the atmosphere, these gases are, in descending order of percentages by volume, N_2 (78.1%), O_2 (20.9%)[2], Ar (0.9%) and CO_2 (0.03%) (the percentages cited here are different from the data in Table 4.2, where they are percentages by mass). The percentage by volume of a gas in the atmosphere corresponds to its

2 In this chapter, O_2 symbolizes a molecule of free oxygen, also called dioxygen.

partial pressure; for example, the partial pressure of O_2 in the atmosphere (pO_2) is 0.209 atm, which means that O_2 accounts for 20.9% of the total atmospheric pressure at sea level. In surface waters of the ocean, the theoretical concentration of each dissolved gas (called "solubility") is equal to the product of its partial pressure in the atmosphere and its solubility coefficient. The latter is different for each gas, and decreases as seawater temperature increases. At 25°C, for example, the solubility coefficient of the four main atmospheric gases is 507 (N_2), 1,019 (O_2), 1,117 (Ar) and 28,400 (CO_2) $\mu mol\ kg^{-1}\ atm^{-1}$ (number of gas molecules dissolved per kilogram of seawater and per unit of atmospheric pressure). It follows that CO_2 is highly soluble in seawater. Hence even if CO_2 only accounts for 0.03% of the volume of the atmosphere, its solubility in seawater ($0.0003 \times 28{,}400 = 9.6\ \mu mol\ kg^{-1}$) is as high as that of Ar ($0.009 \times 1{,}117 = 10.1\ \mu mol\ kg^{-1}$), whereas this gas is 30 times more abundant than CO_2 in the atmosphere (0.9% compared to 0.03%).

We must note that the values in the previous paragraph correspond to the concentrations of chemical elements in seawater in the form of dissolved gases. However, these same chemical elements may also be present in the ocean in other forms, for example hydrogen and oxygen which are extremely abundant in the ocean in the form of water, H_2O. Thus, the total mass of oxygen in seawater (including both dissolved O_2 and H_2O) accounts for 86% of the total mass of this reservoir (Table 4.2), whereas the mass of nitrogen in seawater is a million times smaller.

4.2. A fundamental characteristic of the Earth's system: biogeochemical cycles

The concept of the cycle occurs frequently in science. For example, the phases through which the cells of a living organism pass between two successive divisions is called the cell cycle (Figure 4.2). In the domain of biogeosciences, a biogeochemical cycle refers to the transport and cyclic transformation of a chemical element (e.g. oxygen) or a compound (e.g. CO_2) between the Earth's natural reservoirs cited in section 4.1.1 (i.e. atmosphere, biosphere, geosphere and hydrosphere), as influenced by the activity of living organisms

(Figures 4.3–4.7). During these cycles, the chemical elements may change state several times (i.e. mineral, nonliving organic or living organic) and their compounds may change phase (i.e. solid, viscous, liquid or gaseous).

Different natural cycles can take place at very different spatial and temporal scales. For example, a biochemical cycle may take place within an organelle of a living cell, within a matter of seconds. The cell cycle involves a whole cell or several cells of the same population or tissue, and takes place over several hours or days. Finally, a biogeochemical cycle involves the whole planet and requires much more time to complete than a biochemical cycle or the cell cycle, as explained afterwards. Despite the differences in spatial and temporal scales, all these cycles reflect two fundamental laws of nature. On the one hand, the progressive dissipation of energy in the Universe (i.e. the conversion of different forms of energy – solar, geothermal, nuclear, etc. – into heat) generates flows (movements) of energy and matter, which occur at all spatial scales. On the other hand, as matter "is neither lost nor created", the chemical elements must be recycled at some point or another. The combination of flows and recycling of matter produces cycles at all scales in nature, including biogeochemical cycles.

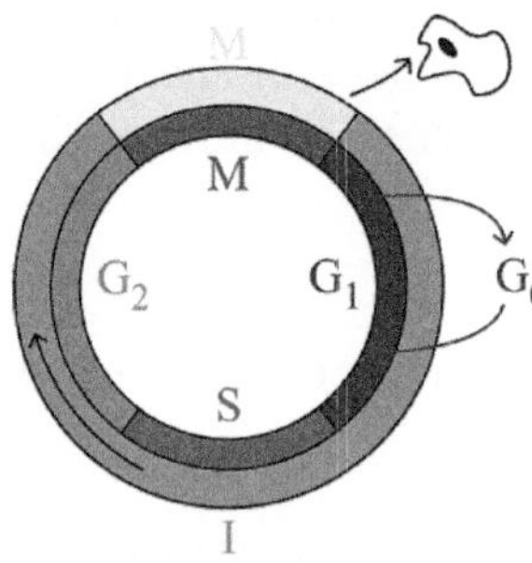

Figure 4.2. *Illustration of the cell cycle whose different phases are: G_0 (Gap 0, resting phase), G_1 (Gap 1, cell growth), S (synthesis, DNA replication), G_2 (Gap 2, cell growth), M (Mitosis, cell division) (from https://upload.wikimedia.org/wikipedia/commons/c/c4/Cell_cycle.png?uselang=fr, original figure downloaded from this site by Magnus Manske) (see color section)*

COMMENTS ON FIGURE 4.2.– After mitosis and a period without division during which the cell grows (G_0 and G_1), the cell can divide again, forming a cycle.

As far as biogeochemical cycles are concerned, all chemical elements naturally present on Earth and many compounds of these elements gradually pass from one reservoir to another and ultimately return to the initial reservoir. Depending on the chemical element or compound considered, the completion of a biogeochemical cycle may take years, decades, centuries, millennia, millions or hundreds of millions of years. The pathways of elements or compounds between natural reservoirs depends on both physical and chemical interactions between the reservoirs and within them the activity of living organisms and ecosystems. This creates a very close interdependence, on the one hand, between the physical and chemical conditions that occur on the Earth's surface and, on the other hand, the activity of living systems. This interdependence, which is a fundamental characteristic of the Earth's system, started when life appeared on the planet, approximately four billion years ago, the Earth having been formed half a billion years earlier. *Earth System Science*, which was founded in the 1990s, advocates the comprehensive study of the Earth as a set of integrated and dynamic systems, using Earth sciences, life sciences, applied sciences and social sciences.

It is not possible to describe all biogeochemical cycles associated with marine ecosystems in a few pages. In this chapter, we will explore the biogeochemical cycles of four rather important chemical elements for marine ecosystems: carbon, oxygen, nitrogen and phosphorus. Table 4.3 shows the distribution of these four elements in different reservoirs of the outer envelopes of the Earth. Other elements strongly involved in marine ecosystems are silicon, sulfur and iron. The elements involved are partly different from those in a text about terrestrial ecosystems where there may be shortages of some resources that are not lacking in the marine environment, e.g. H_2O or potassium (K). As with any rule, there are exceptions, e.g. the desiccation experienced by marine ecosystems of the Mediterranean Sea when that basin almost completely dried up for 1–2 million years when the precursor of the present Strait of Gibraltar closed up in the Messinian period, 6–7 million years ago. Hence, even marine systems can experience shortages of H_2O.

Reservoirs	C	O	N	P
– Atmosphere	7.2×10^2	1.2×10^6	4.0×10^6	
– Ocean	3.9×10^4	1.2×10^9	2.3×10^4	1.0×10^2
– Earth's crust and sediments	1.0×10^8	1.2×10^{10}	2.0×10^8	2.0×10^7
– Biosphere		$<1 \times 10^4$		
- Terrestrial plants	5.6×10^2		3.5×10^0	3.0×10^0
- Soils	1.5×10^3		9.5×10^0	1.0×10^2
– Fossil organic matter	4×10^3			

Table 4.3. *Distribution, in different reservoirs of the outer envelopes of the Earth, of four chemical elements that are discussed in detail in sections 4.3 to 4.6. Each value represents the total mass of the element expressed in Pg C (10^{15} grams) or Gt (10^9 tons). The exploitable part of the phosphorus present in the lithosphere represents only a millionth of the quantity of this element in the reservoir*

Nitrogen, phosphorus, silicon, sulfur and iron are some of the chemical elements required for growth by different types of marine organism. Substances dissolved in seawater containing these essential elements are called inorganic nutrients when their molecules do not contain carbon–hydrogen bonds, and organic nutrients when their molecules contain at least one such bond. Nutrients that are normally present in seawater in micromolar concentration (μmol L^{-1}) are called macronutrients, and those which are usually present in nanomolar concentration (nmol L^{-1}) are called micronutrients.

4.3. Carbon: at the heart of living matter

Carbon is the chemical element at the heart of living matter. The biomass of living organisms and the chemical compounds that come from the decomposition of this biomass form the organic matter, which is distinct from inorganic matter. The area of chemistry that studies carbon compounds is called organic chemistry. A chemical compound is said to be organic when one or more atoms of carbon are linked to atoms of other elements, usually hydrogen, oxygen or nitrogen.

4.3.1. *Carbon in large natural reservoirs*

The carbon cycle involves four large natural reservoirs: the atmosphere, the biosphere (marine and terrestrial living organisms), the hydrosphere (ocean and freshwater) and the lithosphere (crust and sediments, sedimentary rocks and fossil organic matter, i.e. peat, coal and hydrocarbons, including oil and natural gas) (Figure 4.3, Tables 4.3 and 4.4; Table 4.4 details the carbon values given in Table 4.3). Various mechanisms, including the activity of living organisms, move carbon between different inorganic and organic compounds and between the large reservoirs, at different time scales. The following paragraphs briefly describe the different forms of carbon found in the atmosphere, lithosphere, hydrosphere and biosphere.

In the atmosphere, CO_2 represents only a small proportion of gases in this outer envelope of the Earth (<0.1%, Table 4.2), but it greatly contributes to the greenhouse effect (i.e. retention of heat from the Sun in the lower part of the atmosphere). The presence of different greenhouse gases in the atmosphere, including CO_2 and water vapor, keeps the surface temperature of the Earth higher than if they were absent. The greenhouse effect is essential to living systems because, without it, the mean surface temperature on Earth would be –18°C instead of the present +15°C (in fact, at –18°C, ice would cover the globe and the mean temperature would plummet to –50°C). The two most abundant carbon compounds in the atmosphere are carbon dioxide (CO_2) and methane (CH_4). Although CH_4 is almost a hundred times less abundant than CO_2 in the atmosphere, each molecule of CH_4 has an impact 20 times greater than a molecule of CO_2 on the greenhouse effect. However, the lifetime of CH_4 in the atmosphere is only approximately 10 years (CH_4 is transformed into CO_2 and H_2O after several chemical reactions), so its effect on the climate essentially depends on its rate of production. The natural greenhouse effect is therefore an essential condition for the existence of life on Earth. Our current concern about the greenhouse effect is the rapid increase in concentration of atmospheric greenhouse gases due to human activity, which causes the observed increase in temperature of the atmosphere and oceans. This global-scale increase in temperature leads to changes in the environment, the speed of which may exceed the adaptation ability of natural ecosystems and human societies.

Reservoirs	Falkowski *et al.* [FAL 00]	Pidwirny [PID 06]
– Atmosphere	7.2×10^2	5.8×10^2 (in 1700) to 7.7×10^2 (in 1999)
– Oceans (total)	3.8×10^4	3.8 to 4.0×10^4
- Total inorganic	3.7×10^4	
- Total organic	1.0×10^3	
- Surface layer	6.7×10^2	
- Deep layer	3.7×10^4	
– Lithosphere	$>7.5 \times 10^7$	6.6 to 10.0×10^7
- Sedimentary carbonates	$> 6 \times 10^7$	
- Kerogen*	1.5×10^7	
– Terrestrial biosphere (total)	2.0×10^3	2.0 to 2.2×10^3
- Living biomass	6.0 to 10.0×10^2	
- Non-living biomass	1.2×10^3	
- Plants		5.4 to 6.1×10^2
- Soils		1.5 to 1.6×10^3
– Aquatic biosphere	$1–2 \times 10^0$	
– Fossil organic matter (total)	4.1×10^3	4.0×10^3
- Coal	3.5×10^3	
- Oil	2.3×10^2	
- Natural gas	1.4×10^2	
- Other (peat)	2.5×10^2	

*Oil shale and sands, anthracite, graphite, etc.

Table 4.4. *Distribution of carbon in the main reservoirs of the Earth. The values are expressed in Pg C (10^{15} grams de C) or Gt C (10^9 tons of C). Another natural carbon reservoir, not mentioned in the table above, is methane clathrate stored in marine sediments (0.5 to 2.5 × 10^3 Gt C)*

In the lithosphere, there is carbon in sedimentary rocks, which result from the consolidation of marine sediments, in metamorphic rocks and in some magmatic rocks. Marine sediments largely come from organic and inorganic material of biological origin that accumulates on the ocean floor. Over millions of years, sediments are transformed into various rocks including limestone ($CaCO_3$) or into fossil organic matter, including oil and natural gas (whose principal constituent is CH_4). Within the sediments, CH_4 is also present in the form of methane clathrate (also called fire ice), which is a substance where the molecules of CH_4 are trapped in thin cages of ice, the general name of such a chemical structure being clathrate. Most of the methane clathrate is found in deep oceanic sediments, where CH_4 is released through local bacterial transformation of organic sediments (equation [4.14]). A small portion of methane clathrate is found on the continental shelf at depths of hundreds of meters, originating from the migration of CH_4 formed deeper through the thermal decomposition of organic matter. Methane clathrate is younger than hydrocarbons, but the quantity of carbon it represents is similar (see values in Table 4.4). The largest reservoir of carbon on Earth consists of sedimentary rocks.

In the hydrosphere, carbon may be in dissolved or particulate inorganic forms. Dissolved inorganic carbon (DIC) includes the different chemical species of dissolved CO_2 (section 4.3.2). The concentrations of DIC per unit volume of seawater are higher in deep oceanic waters than close to the surface (see Chapter 2 of [MON 14b], section 2.4.1). Particulate inorganic carbon (PIC) is essentially composed of calcium carbonate ($CaCO_3$). It includes calcareous skeletons (also known as "tests") of planktonic organisms (the tests are external structures of organisms that include coccolithophores, foraminifera and pteropods) and external structures of benthic organisms (e.g. corals and shells).

The biosphere includes living organisms, for which part of the biomass is active and part is inert (e.g. the wood of trees). On land, most of the vegetal biomass is inert whereas, in the marine environment, almost all the biomass is active. Hence the terrestrial vegetal biomass has a turnover, on average, of ten years, whereas marine vegetal biomass (phytoplankton) has a turnover, on average, of a day. A second component of inert biomass is made up of particulate and dissolved organic detritus, and the organic fraction of soil and marine sediments. Taking into account this second component, the total inert biomass is much greater than the active biomass in the whole biosphere, including in the ocean.

4.3.2. *Biogeochemical cycles of carbon*

Inorganic and organic carbon have both short and long cycles. Figure 4.3 illustrates the biogeochemical flows of carbon between different natural reservoirs (below, numbers in italics between brackets refer to arrows in the figure). Four carbon cycles are identified in the following text – inorganic and organic, both with short and long cycles. Each cycle consists of a subset of the flows that are represented in Figure 4.3.

4.3.2.1. *Short inorganic carbon cycle*

The short cycle of inorganic carbon begins with the exchange of CO_2 (along with other gases) between the atmosphere and the ocean [*1*], whose direction depends on the partial pressures of this gas (pCO_2) in the two environments. In sea–surface water, dissolved CO_2 reacts with water (H_2O) to form carbonic acid (H_2CO_3), which dissociates into bicarbonate (HCO_3^-) and one hydrogen ion (H^+); bicarbonate can also dissociate into carbonate (CO_3^{2-}) and one hydrogen ion:

$$CO_2 + H_2O \rightarrow H_2CO_3 \rightarrow HCO_3^- + H^+ \rightarrow CO_3^{2-} + 2H^+ \qquad [4.1]$$

Equation [4.1] describes the carbonate system. The arrows in this equation may be reversed, in which case the reaction goes from CO_3^{2-} to CO_2, which may lead to the release of dissolved CO_2 from the ocean into the atmosphere. The direction in which the reaction proceeds, toward the left or the right, strongly depends on the concentration of H^+ ions in seawater and, therefore, on its pH (which is inversely proportional to the concentration of H^+ ions). At the normal pH of seawater, which is approximately 8.2, the different compounds of the carbonate systems have very different abundances:

0.5% of DIC is in the form of dissolved CO_2 and H_2CO_3 (there is much less H_2CO_3 than dissolved CO_2), 89% is in the form of HCO_3^- and 10.5% is in the form of CO_3^{2-}. The sum of all chemical species of DIC is called $\sum CO_2$ or total CO_2:

$$\sum CO_2 = \text{dissolved } CO_2 + H_2CO_3 + HCO_3^- + CO_3^{2-} \qquad [4.2]$$

We explained above (section 4.1.2) that the solubility of atmospheric CO_2 in seawater is much greater than that of other gases such as O_2. Due to this remarkable solubility, a large quantity of atmospheric CO_2 is expected to dissolve in sea–surface water, and the amount of CO_2 that actually dissolves in seawater depends on the strength of the pCO_2 gradient between the two reservoirs. According to conditions, the pCO_2 gradient may be in opposite directions and, in areas where $pCO_{2(ocean)} > pCO_{2(atmosphere)}$, CO_2 is released from the ocean into the atmosphere. Also, due to the reactivity of CO_2 with H_2O, most of the CO_2 in seawater is transformed, at the normal pH of seawater, into HCO_3^- and CO_3^{2-} (equation [4.1]) which stores $\sum CO_2$ in the ocean. As a consequence of the high solubility of CO_2 in seawater and its high reactivity with H_2O, the quantity of $\sum CO_2$ in the ocean is approximately 50 times greater than that of CO_2 in the atmosphere.

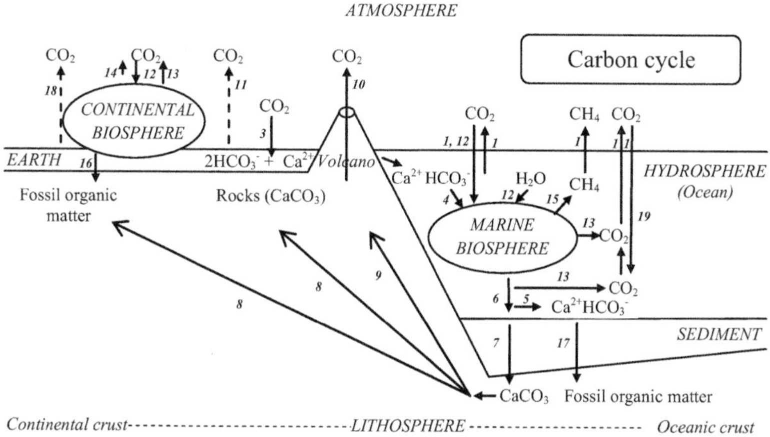

Figure 4.3. *The carbon cycle involves the atmosphere, the biosphere (marine and terrestrial), the hydrosphere (ocean) and the lithosphere (sedimentary rocks and fossil organic matter)*

COMMENTS ON FIGURE 4.3.– The inorganic carbon cycles involve: exchanges of gas between the atmosphere and the ocean, the weathering of continental and marine rocks by atmospheric and dissolved CO_2 (which supplies Ca^{2+} and HCO_3^- to the ocean), the precipitation and dissolution of $CaCO_3$ in the ocean, the tectonic recycling of carbonate sedimentary rocks in volcanos and the calcination of $CaCO_3$ to make cement; the last two processes release CO_2 into the atmosphere. The organic carbon cycles involve: photosynthesis and respiration by the living biomass, oxidation and fermentation of the nonliving biomass in soils and marine sediment, the sedimentation and burial of organic matter in the seabed as well as the weathering of sedimentary rocks, the combustion of fossil organic matter and deforestation, three processes that release CO_2 into the atmosphere. Full arrows represent natural flows and dotted arrows represent flows associated with human activity; the numbers next to the arrows refer to the flow descriptions in the text.

Various physical and biological mechanisms (called "carbon pumps"; see section 2.4.1 in Chapter 2 of [MON 14b]) transport to deep waters some of the inorganic and organic carbon substances that are formed in surface waters. The carbon that reaches deep waters is sequestered there in the form of dissolved CO_2 until these waters return to the surface [*2*] where CO_2 may be released into the atmosphere [*1*]. The physical mechanism of carbon pumping toward deep waters is the downward transport of seawater containing a high concentration of dissolved atmospheric CO_2 [*19*], which is very rapid as it requires only a few months to a few years. This phenomenon occurs in the areas of deep-water formation described in section 2.1.2 in Chapter 2 of [MON 14b]. The return of this CO_2 toward the surface by upwards movements of seawater is much slower, i.e. from several centuries to more than a millennium (see section 2.1.2 in Chapter 2 of [MON 14b]). The biological carbon pumping mechanism involves the export of particles of biological origin that sink toward the seabed [*6*]; this mechanism is described afterwards in the context of the long organic carbon cycle. Since the beginning of the industrial era (1750), the natural pumping of carbon in the ocean has absorbed approximately half the CO_2 released into the atmosphere by human activity (this CO_2 is said to be anthropogenic).

An increase in pCO_2 in the atmosphere causes a rapid increase in the dissolution of CO_2 into the ocean, which causes an increase in the concentration of H^+ ions in seawater as shown in equation [4.1]. Because the pH is inversely proportional to the concentration of H^+ ions, the current increase in atmospheric CO_2 causes a decrease in pH or, in other words, ocean acidification. The sequestration of anthropogenic carbon by the ocean, described in the previous paragraph, is therefore not "free": the price to pay is the acidification of seawater. For more information, refer to the special edition of the journal *Oceanography* dedicated to *Ocean Acidification* [AUT 09]. Note that seawater is not currently acidic (acidity is defined as $pH < 7.0$), since its pH, which is slowly but progressively decreasing, is still usually >8.0.

4.3.2.2. *Long inorganic carbon cycle*

The long inorganic carbon cycle begins with CO_2 dissolving into continental waters [*3*], where it forms H_2CO_3 (see the short cycle, above), which reacts with continental rocks (including limestone, $CaCO_3$) thereby producing calcium ions and bicarbonate:

$$CaCO_3 + H_2CO_3 \rightarrow Ca^{2+} + 2\ HCO_3^- \qquad [4.3]$$

In the sea, ions transported by rivers [*4*] and those already in the ocean may be precipitated by calcification, which is essentially biological in the current ocean:

$$Ca^{2+} + 2\ HCO_3^- \rightarrow CaCO_3 + CO_2 + H_2O \qquad [4.4]$$

This equation summarizes several chemical and biological phenomena. Indeed, at the final stage of $CaCO_3$ precipitation, calcifying organisms combine Ca^{2+} and CO_3^{2-}. However, since CO_3^{2-} is not very abundant in seawater and its transport across biological membranes is not easy, it is often HCO_3^- that is taken from seawater and transported in the organisms (left part of equation [4.4]), where it dissociates into CO_3^{2-} and H^+. The latter H^+ ion chemically combines with an HCO_3^- ion in seawater (i.e. one of two HCO_3^- ions in the left part of equation [4.4]), which generates CO_2 and H_2O (in the right part of equation [4.4]).

The production of organic compounds (here, $CaCO_3$) by biological processes (here, calcification) is called biomineralization, which is a very widespread phenomenon in living organisms (e.g. siliceous frustules of diatoms, carbonate shells of mollusks, coral reefs and bones and teeth of vertebrates). In the ocean, part of the carbonate debris from organisms (PIC, see above) may dissolve in the water column [*5*]. The process of $CaCO_3$ dissolution is the counterpart of calcification:

$$CaCO_3 + CO_2 + H_2O \rightarrow Ca^{2+} + 2\ HCO_3^- \qquad [4.5]$$

The $CaCO_3$ that does not dissolve in the water column is deposited on the seafloor [*6*], where carbonate sediments are slowly transformed into sedimentary rocks [*7*]. Over millions of years, the movements of tectonic plates (orogenesis: formation of mountains) carry part of these rocks to the continents [*8*], where they may be altered by H_2CO_3 (equation [4.3]). Tectonic movements also carry carbonate rocks to volcanoes where they are recycled [*9*], which releases CO_2 into the atmosphere [*10*]. This closes the carbonate cycle.

The biogeochemical carbonate cycle above can regulate the climate in the medium term (time scale of approximately 100,000 years, which corresponds to astronomic and paleoclimatic Milankovitch cycles), but the CO_2 balance of this cycle is null in the longer term. Indeed:

– the formation of one molecule of carbonic acid (H_2CO_3) on continents removes one molecule of CO_2 from the atmosphere (equation [4.1]);

– the weathering of $CaCO_3$ by this H_2CO_3 molecule on land produces two HCO_3^- molecules (equation [4.3]);

– the precipitation of $CaCO_3$ in the ocean combines two molecules of HCO_3^- with one molecule of calcium (Ca^{2+}), producing one molecule of $CaCO_3$, one molecule of water and one molecule of CO_2 (equation [4.4]), which is released to the atmosphere.

Thus, the molecule of CO_2 that is removed from the atmosphere during the weathering of $CaCO_3$ is returned to the atmosphere during the precipitation of $CaCO_3$, so that the net balance of the cycle in the long term is null.

Contrary to the biogeochemical carbonate cycle, the weathering by carbonic acid of rocks that contain silicate minerals can regulate the climate in the long term:

– the formation of two molecules of carbonic (H_2CO_3) removes two molecules of CO_2 from the atmosphere (equation [4.1]);

– the weathering of a silicate mineral on land consumes two molecules of H_2CO_3 and one molecule of water, and produces one molecule of silicic acid, $Si(OH)_4$ (the following example uses wollastonite, a mineral composed of calcium silicate):

$$CaSiO_3 + 2\ H_2CO_3 + H_2O \rightarrow Ca^{2+} + 2\ HCO_3^- + Si(OH)_4 \qquad [4.6]$$

– according to equation [4.6], the reaction of two HCO_3^- molecules with one molecule of calcium (Ca^{2+}), in the ocean, produces one molecule of $CaCO_3$, one molecule of water and one molecule of CO_2, which is released into the atmosphere. Hence for two molecules of CO_2 removed from the atmosphere during the weathering of silicate minerals, one is returned to the atmosphere during the precipitation of $CaCO_3$, so that the net balance of the cycle in the long term is the removal of one CO_2 molecule from the atmosphere.

Equation [4.6] can therefore potentially remove anthropogenic CO_2 from the atmosphere in the long term. However, if this reaction were to proceed with no counterbalance for several hundred thousand years, it would remove all CO_2 from the atmosphere and the Earth, without the greenhouse effect, would freeze (see section 4.3.1). Fortunately, the biogeochemical carbonate-silicate cycle does not end with equation [4.6]. In fact, in the ocean, diatoms (phytoplanktonic algae) use silicon from silicic acid (H_4SiO_4) to form their external envelope (frustule), which is composed of biogenic silicon (called opal, SiO_2):

$$Si(OH)_4 \rightarrow SiO_2 + 2\ H_2O \qquad [4.7]$$

Other planktonic organisms, namely silicoflagellates and radiolaria, also use SiO_2. Siliceous tests of these planktonic organisms sink toward the seabed where they accumulate in sediments, together with $CaCO_3$ debris. The combination of equations [4.1], [4.4], [4.6] and [4.7] can be summarized as follows:

$$CaSiO_3 + CO_2 \rightarrow SiO_2 + CaCO_3 \qquad [4.8]$$

When marine sediments and rocks on the seabed are transported to continents by movements of tectonic plates, the high temperature and pressure in the Earth's mantle combine SiO_2 and $CaCO_3$ into metamorphic rocks:

$$SiO_2 + CaCO_3 \rightarrow CaSiO_3 + CO_2 \qquad [4.9]$$

With equation [4.9] being the opposite of equation [4.8], the biogeochemical carbonate–silicate cycle is closed, which occurs after a very long period of time (i.e. millions of years). On this time scale, the release of CO_2 by volcanos maintains the concentration of this gas in the atmosphere. The biogeochemical carbonate–silicate cycle continues by the progressive uplifting of silicate rocks to the land surface, where they are subjected to weathering by H_2CO_3 (equation [4.6]).

Due to temporal irregularities in tectonic activity on Earth, which includes the formation and fragmentation of supercontinents, equations [4.8] and [4.9] are never in perfect equilibrium. On the time scale of millions of years, the amount of CO_2 in the atmosphere and the climate are governed by the carbonate–silicate biogeochemical cycle, whereas on the time scale of Milankovitch paleoclimatic cycles, they are governed by the biogeochemical carbonate cycle.

Long-term increases in atmospheric temperature are usually accompanied by increases in precipitation. Higher precipitations cause long-term increases in the weathering of rocks, including those that contain silicate minerals (e.g. wollastonite in equation [4.6]). The weathering of silicate rocks causes a net removal of atmospheric CO_2 (equation [4.8]), which leads to a decrease in temperature, which is an example of a global-scale negative feedback.

A major industrial activity in modern societies is the calcination of $CaCO_3$ [*11*] to obtain quick lime (CaO) for the production of cement:

$$CaCO_3 \rightarrow CaO + CO_2 \qquad [4.10]$$

This activity accelerates a small component of the long inorganic carbon cycle by releasing large quantities of CO_2 into the atmosphere at a faster rate than the natural cycle. The production of CO_2 by calcination accounts for approximately 5% of the anthropogenic emissions of this gas and 3% of the total CO_2 emissions, natural and anthropogenic.

4.3.2.3. *Short organic carbon cycle*

The short organic carbon cycle combines various biological processes (also see section 4.4.2). The first of these processes described here is oxygenic photosynthesis [*12*] (for anoxygenic photosynthesis, see section 4.4.1), which uses CO_2 (inorganic), water and energy (in the natural environment, sunlight) to produce organic matter, and releases oxygen and water as by-products (oxygenic means which produces O_2). This very complicated process is simplified in the following equation (in particular, organic matter is represented by the simplest sugar, i.e. glucose):

$$6\ CO_2 + 12\ H_2O + \text{energy (light)} \rightarrow C_6H_{12}O_6\ \text{(sugar)} + 6\ H_2O + 6\ O_2 \quad [4.11]$$

As photosynthesis uses and produces H_2O, equation [4.11] is often replaced by the following, simplified formula:

$$6\ CO_2 + 6\ H_2O + \text{energy (light)} \rightarrow C_6H_{12}O_6\ \text{(sugar)} + 6\ O_2 \quad [4.12]$$

However, equation [4.12] does not clearly show that the oxygen (O_2) produced during photosynthesis comes from (H_2O) and not from CO_2, which is an important aspect of oxygenic photosynthesis for the biogeochemical cycle of oxygen (section 4.4). It has been suggested that diatoms, which are a major component of phytoplankton, are only able to use inorganic carbon in its dissolved gas form (CO_2) excluding the other chemical species of CO_2 in seawater (see above), but it now appears that these organisms can also fix carbon from HCO_3^- [TOR 97]. The concentration of $\sum CO_2$ (equation [4.2]) in sea-surface water is generally considered to be high enough for carbon not to limit phytoplankton photosynthesis [TOR 00], thereby assuming that increases in the current and future concentrations of CO_2 in the atmosphere, and consequently in the ocean, would not directly affect marine primary production. However, experiments on natural plankton assemblages enclosed in mesocosms (large bags of several cubic meters floating in the ocean) artificially kept under three different CO_2 partial pressures showed that the amount of carbon fixed by phytoplankton increased in response to increased pCO_2 [RIE 07]. Hence it is possible that an increase in CO_2 concentration in the ocean could directly affect the functioning of marine ecosystems, but the mechanism responsible for this increase remains unknown.

There is another major process by which organic matter is synthesized in the ocean, i.e. chemosynthesis. In this case, organisms obtain the energy required for the synthesis of organic matter from inorganic chemical compounds, instead of light as in photosynthesis. These organisms are often found in environments without light, for example in hydrothermal systems on the seafloor.

A second group of biological processes involved in the short organic carbon cycle is respiration and aerobic decomposition of organic matter by oxidation [*13*]. With these processes, organisms can use the energy stored in organic matter, with accessory production of CO_2 and water:

$$C_6H_{12}O_6 + 6\ O_2 \rightarrow \text{energy} + 6\ CO_2 + 6\ H_2O \qquad [4.13]$$

In equation [4.13], the oxygen atoms in the H_2O molecules (to the right) come from the O_2 molecules and not from the sugar (to the left), which only supplies the hydrogen atoms. $\sum CO_2$ resulting from the respiration and aerobic decomposition of organic matter in surface waters or the underlying intermediate waters returns to the surface within a few years, from where it may escape into the atmosphere if $pCO_{2(ocean)} > pCO_{2(atmosphere)}$. On the contrary, $\sum CO_2$ that is released into deep ocean waters may be sequestered there for several centuries, within the deep ocean circulation (see section 2.1.2 in Chapter 2 of [MON 14b]).

Equation [4.13] also describes the combustion (i.e. burning) of organic matter, where the energy produced is heat. In fact, respiration is a specific type of combustion where metabolic cycles or processes of living organisms strictly control the release of energy from organic matter, whereas burning is not naturally controlled. Because most of the vegetation biomass removed from forests is burnt in the short or medium term, the human activity of deforestation [*14*] accelerates the transfer of carbon from the continental biosphere to the atmosphere.

Two types of anaerobic respiration are mentioned later (section 4.4.2), namely nitrate respiration or denitrification and sulfate reduction (or respiration) (arrow [*14*] in Figure 4.4). Denitrification is explained in more detail in section 4.5.2.

A third group of biological processes are involved in the short organic carbon cycle, i.e. fermentation, which takes place in an anaerobic environment (i.e. in the absence of O_2). In the example below, fermentation uses the energy contained in organic matter, and produces two inorganic substances, i.e. CO_2 and CH_4 [*15*]:

$$C_6H_{12}O_6 \rightarrow 3\ CO_2 + 3\ CH_4 + \text{energy} \qquad [4.14]$$

There are several types of fermentation. For example, during the production of alcoholic drinks, fermentation produces two molecules of ethanol (C_2H_5OH) and two molecules of CO_2 from one molecule of glucose ($C_6H_{12}O_6$). In the water column, fermentation takes place in the digestive tract (anaerobic environment) of various organisms, including zooplankton and fish, where it contributes to food digestion.

The different processes above are coupled with variations in atmospheric CO_2 and CH_4 on time scales of several years to decades. Organisms that synthesize organic matter from organic compounds are called autotrophs and those which feed on existing organic matter are called heterotrophs; some organisms, called mixotrophs, are capable of using the two different sources for nutrition. All organisms – autotrophs, heterotrophs or mixotrophs – respire.

4.3.2.4. *Long organic carbon cycle*

The long organic carbon cycle links the photosynthetic production of organic matter [*12*], its export to the ocean depths [*6*] and its burial in soils [*16*] and in marine sediments [*17*] where it is slowly transformed into fossil organic matter (including oil and natural gas, accumulated or not in deposits). The processes of hydrocarbon deposit formation and their natural oxidation during transfer to continents by tectonic movements [*8*] correspond to a time scale of hundreds of millions of years (which is the lifetime of the sea floor). As with the calcination of $CaCO_3$ mentioned above [*11*], the evermore intense exploitation of fossil organic matter deposits (coal and hydrocarbons) by humans since the beginning of the industrial era (year 1750) [*18*] shortens part of the long carbon cycle by releasing large quantities of CO_2 into the atmosphere, at a much faster rate than the natural cycle. This human activity is responsible for the majority of anthropogenic CO_2 emissions. As the remainder of the long organic carbon cycle remains unchanged in the short-term, there is a transitory increase in CO_2 in the terrestrial system. It must be noted that getting access to fossil organic matter has been the cause of numerous armed conflicts between countries for over a century.

The brief summary above shows that the long and short cycles of inorganic and organic carbon are strongly interconnected. This makes predicting the exact effects of the current increase in atmospheric CO_2 on our planet over the next decades incredibly difficult (although the longer-term global effect is known). Hence the climate models that have been developed take into account an increasing number of interactions between the different forms of carbon and between the different reservoirs on Earth.

4.4. Oxygen: a poison that Earth cannot do without

Oxygen is the most abundant element in the Earth's crust, where it is present in oxides of silicon, aluminum and iron, as well as in the ocean, where it is both the main component of water molecules and a major dissolved gas (Table 4.2). An original characteristic of our planet is the presence of free elemental oxygen in the atmosphere and hydrosphere, i.e. molecular O_2. In fact, oxygen does not usually exist as a free chemical element in the Universe because it is extremely reactive and easily combines with other elements to form oxides (water is an oxide of hydrogen). One exception to this very general rule is planet Earth, where O_2 is constantly resupplied by oxygenic photosynthesis, of which the O_2 molecule is a byproduct (equation [4.11]). The presence of O_2 in the Earth's atmosphere depends on the imbalance that exists between photosynthesis and aerobic respiration on the global scale, which is controlled by the deposition of organic materials in marine sediments and soils where they cannot be oxidized. This stock of organic matter largely determines the quantity of O_2 in the atmosphere, as it is made up of organic matter coming from reaction [4.11] which was not destroyed by reaction [4.13].

Table 4.3 shows that most of the oxygen in the Earth's outer envelopes is found in lithospheric minerals (10^{10} Pg of oxides in the crust and mantle) and in the ocean (10^9 Pg). In the latter, oxygen is largely present in the form of water molecules (H_2O; > 97% of water on the planet is found in oceans) and, secondly, in the form of dissolved gas (O_2). The atmosphere only contains a small fraction of the oxygen on Earth (10^6 Pg) and the biosphere, even less ($<10^4$ Pg).

In section 4.3.2 above, it was explained that oxygen (O_2) produced during photosynthesis comes from the water molecule (H_2O) (equation [4.11]), and that the oxygen atom in the water molecule produced during respiration comes from O_2 (equation [4.13]). It follows that H_2O forms an integral part of the oxygen cycle. Yet, curiously enough, several geochemistry manuals do not mention H_2O in their chapter on oxygen, forgetting that the ocean contains a large part of the Earth's oxygen in the form of H_2O. In these manuals, the discussion of oxygen in the ocean is limited to just the gaseous O_2 dissolved in seawater. The oxygen cycle in Figure 4.4, below, takes into account H_2O, as it should be.

4.4.1. *The Great Oxygenation Event*

All eukaryotic photosynthetic organisms (eukaryotes are organisms whose cells possess a nucleus; photosynthetic eukaryotes include plants and algae) use oxygenic photosynthesis (which generates O_2), whereas almost all prokaryotic photosynthetic organisms (prokaryotes are organisms whose cells do not possess a nucleus, e.g. bacteria) use anoxygenic photosynthesis (which does not generate O_2). A single group of prokaryotes uses oxygenic photosynthesis, i.e. the cyanobacteria, which are highly abundant in the ocean. Hence the absence of O_2 in the atmosphere of a planet would not necessarily mean that this planet is devoid of life, since life there could be anoxygenic (as was the case on Earth for more than a billion years), but the presence of O_2 in the atmosphere of this planet would be an unequivocal indication of the presence of life.

However, O_2 is a poison to living organisms. It is a deadly poison to strict or obligatory anaerobic organisms, such as certain bacteria that cannot tolerate the presence of any oxygen. At high concentrations, O_2 is a violent poison even for aerobic organisms (who require oxygen for respiration, for example humans) and also for facultative anaerobic bacteria (which prefer oxygen to be present, but can live without it). At low concentrations, O_2 is a slow poison for all organisms using it. The oxygen that we breathe is essential for our survival, but it gradually causes our decline since reactive oxygen species (often called free radicals) are among the causes of ageing.

Oxygenic photosynthesis (simply called photosynthesis from here) was "invented" by aquatic cyanobacteria three billion years ago. Before then, only anaerobic organisms existed on Earth and, when O_2 began increasing in the atmosphere, they could only survive in places without oxygen. Today, we find anaerobic organisms in marine sediments, in aquatic zones devoid of oxygen and, which may seem unlikely, sheltered inside the bodies of aerobic organisms. For example, there are a thousand times more anaerobic bacteria than aerobic bacteria in the digestive tract of human beings, where the anaerobic process of fermentation (see section 4.3.2) contributes to food digestion. Moreover, different forms of fermentation have been domesticated by humans, for the production of cheese, yogurt, alcoholic drinks, sauerkraut, sausage, bread, products derived from soy beans (including soy sauce and tofu) and many other food products for which fermentation helps prolong preservation.

For 2.5 billion years after the appearance of photosynthesis on Earth, organic matter gradually accumulated in deep marine sediments, where it was protected from decomposition (oxidation). As O_2 was not used in the oxidation of this organic matter, this gas accumulated in the atmosphere until it reached a concentration that was toxic to anaerobic organisms. They therefore suffered mass extinction, except those which survived in oxygen-free environments (see previous paragraph). This is known as the Great Oxygenation Event.

Half a billion years after photosynthesis appeared on Earth, aerobic organisms began to develop, i.e. organisms that used O_2 for respiration. Today, aerobic respiration (see section 4.3.2) is used by a wide variety of living beings, from photosynthetic organisms to humans. Their respiration, combined with changes in the dynamics of oceans (whose exact nature is still under debate), progressively stabilized, half a billion years ago, the oxygen concentration in the atmosphere to its current value (forms of respiration other than that of O_2 will be explored in section 4.4.2).

4.4.2. *Biogeochemical cycles of oxygen*

Figure 4.4 shows the biogeochemical flows of oxygen between natural reservoirs (below, numbers in italics between brackets refer to arrows in the figure). Two oxygen cycles are presented here: the first describes the circulation of this element between the atmosphere, the biosphere and the lithosphere and the second shows reactions that take place in the upper atmosphere. Each cycle is composed of a subset of flows as illustrated in the figure. Several of the oxygen flows mentioned below were described in section 4.3.2 since the circulation of oxygen is often coupled with that of carbon.

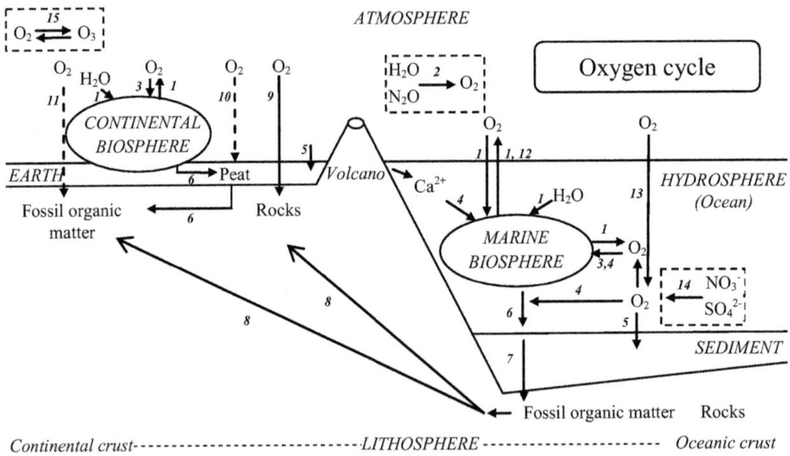

Figure 4.4. *In the oxygen cycle, O_2 circulates constantly between the atmosphere, biosphere and lithosphere (the Earth's crust and mantle)*

COMMENTS ON FIGURE 4.4.– The main driver of the cycle is the production of O_2 by photosynthesis and its use in the respiration, decomposition and combustion (oxidation) of organic matter. Another cycle takes place in the upper atmosphere, where ultraviolet radiation from the Sun constantly transforms molecules of O_2 into ozone (O_3) and molecules of O_3 into O_2. Full arrows represent natural flows and dotted arrow represent flows associated with human activity; the numbers next to the arrows refer to the flow descriptions in the text.

The biosphere produces, through photosynthesis, almost all of the O_2 in the atmosphere by splitting the H_2O molecule [*1*]. During that process (equation [4.11]), photosynthetic organisms separate the hydrogen and oxygen atoms contained in the water molecule (H_2O); they use the hydrogen atoms to produce organic matter and release oxygen atoms in the form of O_2. At least half of this O_2 is produced in the ocean. In addition to photosynthesis, a small amount of O_2 is produced by the photolysis (i.e. decomposition) of water molecules (H_2O) and nitrogen oxide (N_2O) in the atmosphere by the ultraviolet radiation from the Sun [*2*]. The amount of O_2 produced by photolysis is less than 0.01% the amount of O_2 produced by photosynthesis. The biosphere is not only the main producer of O_2, but it is also its main user, for respiration (all aerobic organisms) [*3*] and the decomposition of organic matter (heterotrophic bacteria) in the water column [*4*] as well as in the sediments and soil [*5*]. The oxygen cycle is coupled with both the carbon and water cycles (Figure 4.5).

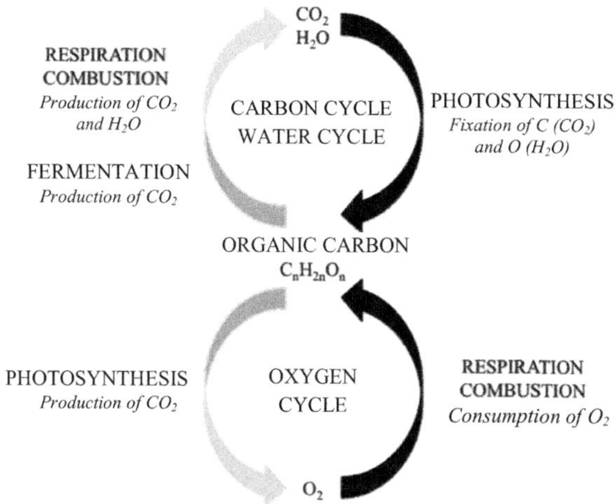

Figure 4.5. *The biological processes of photosynthesis (oxygenic) and respiration (aerobic) and the physico-chemical process of combustion couple the carbon cycle with the water and oxygen cycles. Fermentation also couples the carbon cycle with these other two cycles (adapted from a figure published on the site "Biologie et Multimédia" of University Pierre and Marie Curie (www.snv.jussieu.fr/bmedia/Photosynthese-cours/01-organisms.htm) with permission from UFR 927 "Sciences de la Vie")*

The burial of organic material in the lithosphere [*6*] may prevent it from being combusted, respired or decomposed for very long periods of time. For example, peat (formed in the soil) may exist for thousands of years and its products of transformation (including lignite, coal, bituminous coal and anthracite) for millions of years to hundreds of millions of years. Part of the fossil organic matter that accumulates in the marine sediments [*7*] is transported toward continents by tectonic plate movements (orogensis) [*8*]. Magmatic rocks (basalt) form the substrate on which marine sediments deposit (oceanic plates) [*8*]. On land, different processes consume O_2, including the natural weathering (oxidation) of rocks [*9*] as well as the combustion of peat and [*10*] fossil organic matter [*11*] by human societies

In the ocean, the concentration of O_2 dissolved in the surface layer is usually high due to exchanges with the atmosphere (which occur in both directions) [*12*] and phytoplankton photosynthesis [*1*]. This is also the case at depth due to the thermohaline circulation (see section 2.1.2 in Chapter 2 of [MON 14b]) that transports oxygen from surface waters to ocean depths [*13*] (the vertical movements of water that transport oxygen from the ocean surface toward intermediate and deep waters are described in section 2.1.2 in Chapter 2 of [MON 14b]). Between the surface and deep waters, there is often a minimum in oxygen concentration, at least at low and mid latitudes. This minimum is located at depths where the respiration and decomposition of organic matter consume O_2 faster than its replenishment from the surface. In some cases, the concentration of O_2 at intermediate depths is very low, in which case oceanographers use the expression oxygen minimum zone (OMZ; see next paragraph).

Ocean deoxygenation is an important, and still little known phenomenon even by oceanographers [KEE 10]. This phenomenon refers to the decrease in O_2 in the open-ocean OMZs, which must not be confused with local losses of oxygen in shallow coastal waters incorrectly known as dead zones (see section 4.5.2). Over the last half-century since the 1960s, the concentration of O_2 between 300 and 700 m depths has greatly decreased in several regions of the ocean. This phenomenon is likely due to the increase in surface water temperature, which is associated with global warming. Indeed, higher surface temperature causes both a decrease in the amount of atmospheric gas that can dissolved in the seawater (the solubility of gases, including O_2, is inversely proportional to the temperature of water) and an increase in the density gradient between surface waters – whose density decreases as they warm up – and underlying waters – which are cooler than those in the surface layer, and hence, denser; increased density stratification reduces vertical mixing, and thus, the O_2 replenishment of intermediate waters. What does the future hold? It appears that, as global warming proceeds, the O_2 content of intermediate waters will continue to decrease throughout the world's oceans. The deep ocean would experience a decrease in O_2 of 20–40%, without creating large oxygen-depleted zones. It follows that the water layer most threatened by ocean deoxygenation would be that of intermediate waters, which have been little studied until now.

When the concentration of O_2 is low, some anaerobic bacteria use the oxygen contained in nitrate (NO_3^-) [*14*], which transforms this inorganic nitrogenous nutrient into a dissolved gas (N_2) by a process called nitrate respiration or denitrification (section 4.5.2). When the concentration of O_2 is null, other bacteria, which are strictly anaerobic, use the oxygen contained in sulfate (SO_4^{2}-) [*14*], a process called sulfate reduction (or respiration), which produces hydrogen sulfide (H_2S), a toxic gas with the characteristic foul odor of rotten eggs. In the sea, the production of H_2S primarily takes place in sediments or in deep anoxic zones (i.e. zones without oxygen). In the Black Sea, which is an exceptional case where density stratification of the water column prevents periodic replenishment of O_2 at depth (see section 2.1.2 in Chapter 2 of [MON 14b], most waters below 100–200 m have no O_2 and high H_2S content. Most of the H_2S currently present in the atmosphere is a by-product of the refining and combustion of fossil organic matter by human activities.

Another cycle that involves oxygen is the production and destruction of ozone (O_3) in the upper atmosphere, i.e. 20 to 30 km above the ground [*15*]. There, the ultraviolet radiation from the sun splits some O_2 molecules into individual oxygen atoms (O), which react with O_2 to give O_3:

$$O_2 + O \rightarrow O_3 \quad [4.15]$$

The ozone molecule is destroyed when it combines with an oxygen atom, producing two molecules of O_2:

$$O_3 + O \rightarrow 2\,O_2 \quad [4.16]$$

In the atmosphere, the gases O_2 and N_2 absorb the ultraviolet radiation (UV) from the Sun of short wavelengths (<200 nm). In the upper atmosphere, the presence of the ozone layer extends the UV absorption to longer wavelengths, including UV-C (UV radiation whose wavelengths are between 200 and 280 nm) and most UV-B (UV radiation whose wavelengths are between 280 and 315 nm). The remarkable absorption power of the ozone layer is shown by the fact that UV-B radiation at ground level is 350 million times weaker than that above the ozone layer in the upper atmosphere. Even with that natural protection, the UV-B radiation that reaches the Earth's surface continually damages the DNA of living organisms, which causes cancers and other cell abnormalities when damage occurs at a rate that is faster than the cells' abilities to repair their DNA. Without the UV protection provided by the ozone layer, most organisms that currently live on the surface of the Earth could not survive there.

4.5. Nitrogen: a chemical element over which countries have fought in the past

Nitrogen is an essential constituent of amino acids and proteins. Amino acids themselves are major constituents of proteins, which determine the structure and function of living organisms. Many proteins are enzymes, which catalyze biochemical metabolic reactions. Nitrogen is also a constituent of nucleic acids (RNA and DNA), which are the most complex molecules that exist in nature. Nucleic acids carry instructions that encode all biological molecules and transmit genetic information from cell to cell during cell division (Figure 4.2). Moreover, nitrogen is present in chlorophyll, the pigment of photosynthesis (sections 4.3 and 4.4). Hence nitrogen is an essential chemical element for life on Earth and a key element in biogeochemical cycles.

4.5.1. *Nitrogen, abundant but difficult to access*

On Earth, nitrogen is found in many natural reservoirs. It is much less abundant than carbon or oxygen in the lithosphere, ocean and biosphere, but much more abundant than these two chemical elements in the atmosphere (Table 4.3). In fact, gaseous nitrogen (N_2) accounts for 75% of the mass and almost 80% of the volume of the Earth's atmosphere (Table 4.2 and section 4.1.2). Given the abundance of this chemical element in the atmosphere, one could assume that living organisms have easy access to it. This is, however, not the case, as explained afterwards in this section.

Despite the high abundance of gaseous nitrogen (N_2) in the environment, most living organisms cannot directly fix this form of nitrogen. Indeed, most autotrophic organisms must use nitrate (NO_3^-) or ammonium (NH_4^+) as sources of nitrogen, which chemical forms may be lacking in certain areas or at certain times of the year, and most heterotrophic organisms must obtain their nitrogen from organic matter, which is not always available in sufficient quantities. Hence despite the abundance of N_2, nitrogen may limit the development of many living organisms and ecosystems, since the forms of nitrogen they need may be in short supply.

As mentioned above, nitrogen is the most abundant constituent of the atmosphere. The abundance of N relative to other chemical elements in other reservoirs than the atmosphere is as follows:

– in sea-surface waters, nitrogen is the most abundant dissolved gas since the solubility in seawater of the four main atmospheric gases at 25°C is 383 (N_2), 207 (O_2), 10.1 (Ar) and 9.6 (CO_2) µmol kg^{-1}. However, the total mass of nitrogen in seawater is much lower than that of hydrogen or oxygen because these two chemical elements make up the water molecules and are therefore the most abundant in the ocean (Table 4.2 and section 4.1.2);

– in living matter, nitrogen is the fourth most abundant chemical element (expressed in terms atoms). For example, the four most abundant chemical elements in the human body (as percentages of the total number of atoms) are: H (63%), O (24%), C (12%) and N (1%). Another example is provided by marine plankton, where the atomic ratio O:C:N (the ratio does not take H into account, this element being the most abundant in atoms) is 13.3:6.6:1.0 (these values are known as the Redfield ratio);

– in deep-ocean waters and marine sediments, organic matter is decomposed by heterotrophic bacteria. This releases into environment nitrogen contained in organic matter, primarily in the form of nitrogenous nutrients. The same phenomenon occurs in soil.

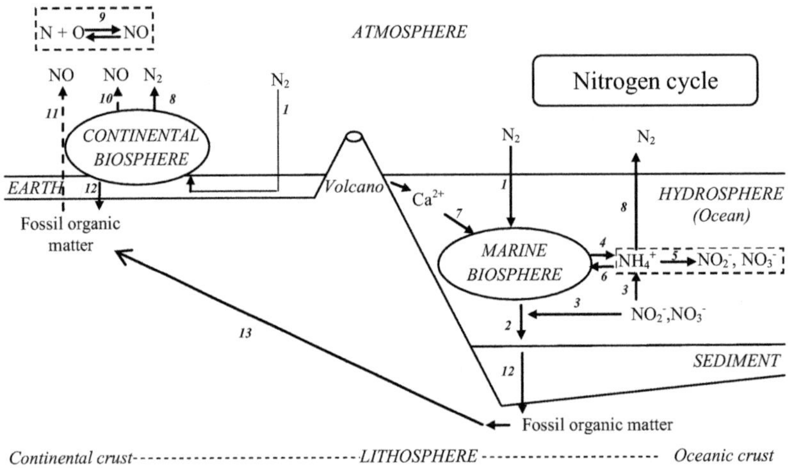

Figure 4.6. *The nitrogen cycle occurs in a similar fashion in the sea and on land*

COMMENTS ON FIGURE 4.6.– The main steps of this cycle are: N_2 fixation by specialized bacteria, the regeneration of nitrogenous nutrients in deep waters, the production of ammonium (NH_4^+) by ammonification, the nitrification of NH_4^+ into NO_2^- and NO_3^- by bacteria, the use of nitrogenous nutrients and denitrification, primarily by denitrifying bacteria, and the process of anammox (see the text). The cycle also includes N_2 fixation by lightening in the atmosphere, which produces NO, and the combustion of wood and fossil organic matter (coal and hydrocarbons), which also produces NO and other nitrogen oxides (NO_x). Full arrows represent natural flows and dotted arrows represent flows associated with human activity; the numbers next to the arrows refer to the flow descriptions in the text.

4.5.2. *Biogeochemical cycles of nitrogen*

Figure 4.6 depicts the biogeochemical flows of nitrogen between natural reservoirs (below, numbers in italics between brackets refer to arrows in the figure). These flows form the nitrogen cycle, which is similar on land and in the sea, and includes the main stages described below.

Fixation of N_2 (gas) by specialized cyanobacteria [*1*], which produces ammonium (NH_4^+):

$$2N_2\ (\text{gas}) + 3\ CH_2O + 3\ H_2O + 4\ H^+ \rightarrow 4\ NH_4^+ + 3\ CO_2 \quad [4.17]$$

where CH_2O represents organic matter. This reaction is catalyzed by the enzyme nitrogenase. The NH_4+ is used to synthesize organic matter, which brings nitrogen of atmospheric origin into the food web. In the sea, the most remarkable N_2 fixing bacteria are cyanobacteria of the genus *Trichodesmium*, which can form filaments of a several cells, which in turn can join to form colonies with diameters of 1 to 10 mm. Filaments or colonies of *Trichodesmium* may span tens or even thousands of square kilometers at the surface of warm seas. In the soil, N_2 fixing cyanobacteria are frequently found in symbiosis with plants, in particular in the roots of leguminous plants.

The presence of N_2 fixing bacteria in the roots of leguminous plants explains why farmers traditionally alternate between crops of legumes, cereals, oleaginous plants and other varieties (called crop rotation), which periodically enriches the soil with symbiotic nitrogen. Today, this practice has been replaced by the addition of synthetic nitrogenous fertilizers to soils, although some countries encourage returning to crop rotation. Artificial fertilizers have themselves replaced previously used natural fertilizers consisting of guano, i.e. excrements of marine birds and bats, rich in nitrogen and phosphorus. In order to understand the importance of large guano deposits prior to the advent of artificial fertilizers, one must remember that the ownership of such deposits caused wars between Chile, Bolivia and Peru in the 19th Century. Due to its key role in global agriculture, guano was the oil of that era.

From the ocean surface, organic matter continually sinks toward deep waters, hence the name marine snow [*2*]. Nitrogen contained in the sinking organic matter is regenerated at depth (see the relevant stages of ammonification and nitrification, below) and nutrients thus produced are returned to the surface by upward movements of water

[*3*] (see section 2.1.2 in Chapter 2 of [MON 14b]). On land, nitrogen is largely regenerated in soils.

Production of NH_{4+} by ammonification [*4*]. The organic compounds synthesized by autotrophs circulate in the aquatic and terrestrial food webs, where several heterotrophic organisms excrete nitrogenous compounds, including ammonia (NH_3; aquatic invertebrates, bony fish and amphibians), urea (cartilaginous fish, including sharks, land amphibians and most mammals) and uric acid (insects, reptiles and birds). Nitrogen compounds of metabolic origin as well as from organic detritus are transformed into NH_4^+ by a process called ammonification.

Nitrification by bacteria in water or soil [*5*]. During the next stage, called nitrification, NH_4^+ is transformed into nitrite (NO_2^-) and nitrate (NO_3^-) by bacteria that live in water or soil:

$$2\ NH_4 + 3\ O_2 \rightarrow 2\ NO_2^- + 2\ H_2O + 4\ H^+ \qquad [4.18]$$

$$2\ NO_2^- + O_2 \rightarrow 2\ NO_3^- \qquad [4.19]$$

Use of nitrogenous nutrients [*6*]. Nitrogenous macronutrients (NO_3^-, NO_2^- and NH_4^+) are used by autotrophic organisms (e.g. marine phytoplankton in sea-surface waters) for the synthesis of new organic matter. In the sea or on land, the rate of production of autotrophic organisms (phytoplankton, vegetation) strongly depends on the rate of replenishment of nitrogenous nutrients.

In the ocean, when the rate of replenishment of nitrogenous nutrients is low, especially when stratification of the water column prevents or slows down the return to the surface of nutrients that have been regenerated at depth [*3*], autotrophic production is low (for example, in the center of the large ocean gyres or during the summer in temperate regions). In contrast, when the rate of replenishment of nitrogenous nutrients is high, for example in coastal regions that receive inputs from rivers flowing through areas of intensive agriculture and therefore transporting large amounts of NO_3^- [*7*], the production of phytoplanktonic or benthic algae may be so high that the ecosystem cannot use it all. The organic matter that is not used decomposes locally, thus consuming the O_2 dissolved in the water. Such excessive nutrient enrichment is called eutrophication, the resulting phytoplankton bloom often colors the water and is called green or red tide (or another color depending on the type of phytoplankton), and the accompanying large decrease in the concentration of dissolved O_2 is qualified, according to the intensity of the phenomenon, of hypoxia (O_2 deficiency) or anoxia (absence of O_2). Waters with low or no O_2 are often called "dead zones", which is incorrect since anaerobic organisms can thrive under anoxic conditions.

In between conditions of extreme nutrient deficiency and those of excessive nutrient richness, described in the previous paragraph, the availability of nitrogenous nutrients in sea-surface waters of temperate or polar regions at the end of winter or during spring often cause phytoplankton blooms, which may be grazed by zooplankton, which are in turn consumed by fish and other large marine organisms (mammals, seabirds, etc.). The abundance of nutrients in these regions at that time of the year is explained by the vertical mixing of the water column that has taken place in the previous months, bringing to the surface nutrients that had been regenerated in deep waters. Even when nutrients are abundant, the phytoplankton bloom cannot occur as long as photosynthesis is limited by light, e.g. due to low solar radiation in winter, or strong vertical mixing of the water column that transports phytoplankton to depth where there is little light. The seasonal increase in light and stratification provides favorable conditions for phytoplankton to bloom, which rapidly uses up the surface nutrients, which explains why blooms are usually of short durations. During the following summer period, stratification of the water column often prevents or reduces the transport of nutrients to the surface, which limits phytoplankton growth despite sufficient solar radiation. In those coastal zones where upwelling of deep water is frequent or permanent, and commercial fisheries are very prosperous, e.g in the upwelling zones of Mauritania, Namibia, Chile and California.

Denitrification and anammox. The nitrogen cycle is closed by the transformation of NO_3^- into gaseous N_2, primarily by the action of specialized bacteria – called denitrifying bacteria – that are found in both water and soil. This process (also called nitrate respiration) only occurs at low oxygen concentration, and is therefore found in OMZs (section 4.4.2):

$$4\ NO_3^- + 5\ CH_2O + 4\ H^+ \rightarrow 2\ N_2\ (gas) + 5\ CO_2 + 7\ H_2O \quad [4.20]$$

where CH_2O represents organic matter. In the ocean, an additional process, called anaerobic ammonium oxidation (anammox) transforms NO_2^- and NH_4^+ into N_2:

$$NO_2^- + NH_4^+ \rightarrow N_2 + 2\ H_2O \quad [4.21]$$

The N_2 produced by these two processes is returned to the atmosphere [*8*], where the nitrogen cycle starts again.

In addition to N_2 fixation by bacteria [*1*], other processes can fix nitrogen. These include the electrical discharges that occur naturally in the atmosphere (i.e. lightning), which combines atoms of N and O to form nitrogen monoxide (NO) [*9*]:

$$N_2 + O_2 \rightarrow 2\ NO \quad [4.22]$$

Moreover, human activities that include the combustion of wood [*10*] and fossil organic matter (coal and hydrocarbons) [*11*] produce NO and other nitrogen oxides (NO_x), which are very toxic gases that contribute to the greenhouse effect. Coal is formed on land [*12*], whereas hydrocarbons are formed at the bottom of the sea from ocean sediments [*12*], from where they are transported onto land by tectonic plate movements (orogenesis) [*13*].

The two opposite processes of N_2 fixation and denitrification are probably in equilibrium at the global scale over time periods of several centuries or millennia. It is not certain whether these processes are in equilibrium locally or regionally over times periods of years or decades and whether they can rapidly respond to the current anthropogenic changes. In this respect, if N_2 fixation stopped in the ocean (which is not considered possible), all the available nitrogenous marine nutrients would be used up within 2,000 years (i.e. a very short period of time geologically speaking), which would put an end to marine biological production.

In the ocean, oceanographers distinguish between new nitrogen and nitrogen regenerated in the part of the water column where photosynthesis creates new organic matter (i.e. in surface waters). New N is the nitrogen that enters surface waters from external sources, by vertical mixing (mainly NO_3^-) and by other processes such as N_2 fixation. Regenerated N is the nitrogen that is recycled in surface waters (mainly NH_4^+). The synthesis of organic matter by phytoplankton is called primary production (section 4.1.2); new nitrogen gives rise to new primary production, and regenerated nitrogen to regenerated primary production. The rate at which organic matter can be exported from a pelagic marine ecosystem, for example toward deep waters or by fishing, cannot exceed its new production otherwise the ecosystem would disappear. In an oceanic area where all nitrogen would come from recycling within surface waters (i.e. an area where new production would be zero), no organic matter could be exported from the surface layer, for example by fishing, without destroying the ecosystem. Compared with economics, regenerated production would correspond to capital and new production, to interest. Hence, the sustainable exploitation of marine renewable resources requires that the maximum amount of biomass that is taken per unit time is not greater than new production. In other words, sustainability requires that the renewal of the exploited biomass per unit time be equal to or greater than regenerated production. This is a necessary but insufficient condition because the equilibrium of an ecosystem may still be disturbed despite a rate of removal of biomass that is smaller than new production, for example when there is overfishing of species or groups of species that play a key role in structuring the food web.

4.6. Phosphorus: a chemical element over which countries may fight in future

In living organisms, phosphorus (P) is an essential constituent of the nucleic acids DNA and RNA (section 4.5.1), adenosine triphosphate (ATP), which provides the energy required for metabolic chemical reactions, and phospholipids, which form cell membranes. Phosphorus also contributes to strengthening bones and teeth, which contain 80–85% of the P in the human body.

4.6.1. *Phosphorus, not very abundant or exploitable*

Unlike the carbon, oxygen and nitrogen cycles described above (sections 4.3–4.5), the phosphorus cycle does not involve the atmosphere since P does not exist in gaseous state in significant amounts in the Earth's atmosphere. Large reservoirs of inorganic P include the lithosphere (rocks that contain apatite, a phosphate mineral), soils and marine sediments, the hydrosphere and the biosphere (Table 4.3). It must be noted that the amount of exploitable phosphorus present in the lithosphere is relatively small, i.e. only a millionth of the phosphorus that exists in that reservoir; this crucial point will be discussed at the end of the next section.

In marine plankton, the atomic ratio O:C:N:P (the ratio does not take H into account, this element being the most abundant in atoms) is 212:106:16:1. These values are known as the Redfield ratio, as mentioned in section 4.5.1 for the O:C:N ratio.

4.6.2. *Biogeochemical cycles of phosphorus*

There are two phosphorus cycles, i.e. the short cycle, which is essentially organic, and the long cycle, which is inorganic. The short phosphorus cycle is embedded in the long cycle. Figure 4.7 shows the biogeochemical P flows between natural reservoirs (below, numbers in italics between brackets refer to arrows in the figure). The short and the long phosphorus cycles are made of subsets of these flows.

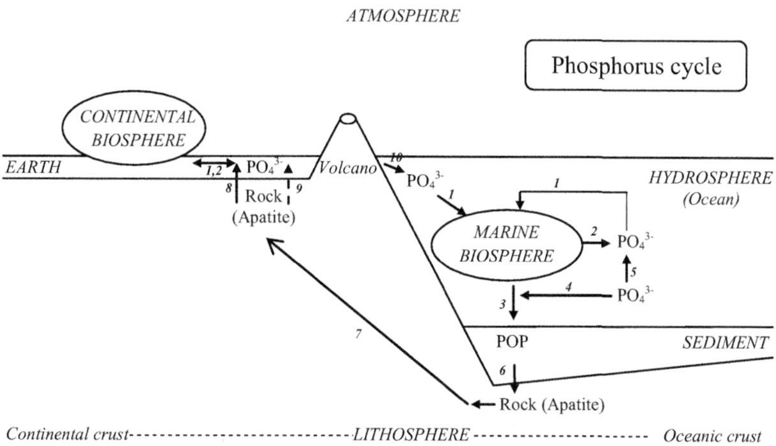

Figure 4.7. *The phosphorus cycle does not involve the atmosphere. Large reservoirs of inorganic P include the lithosphere, soil and marine sediments, the hydrosphere and the biosphere. The short P cycle is essentially organic and the long cycle is inorganic, the first being interconnected with the second. Full arrows represent natural flows and dotted arrows represent flows associated with human activity; the numbers next to the arrows refer to the flow descriptions in the text*

The duration of the short phosphorus cycle ranges between a few days (sometimes less) and a few years. In the soil or water, phosphorus is available to living organisms as the macronutrient phosphate (PO_4^{3-}, of which there are many chemical forms in the natural environment). However, there is usually little PO_4^{3-} in water or soil, and hence, the little bioavailable P is rapidly taken up by photosynthetic organisms [*1*]. Moreover, the latter must often compete for PO_4^{3-} with heterotrophic bacteria. The P that is incorporated in organic matter rapidly circulates in the food web and is finally excreted by metabolic activity or is recirculated by bacterial decomposition of organic matter [*2*]. In the ocean, part of the particulate organic phosphorus (POP) that has sunk to depth (marine snow) [*3*] is regenerated in deep waters to PO_4^{3-} [*4*], which returns to the surface during episodes of vertical mixing or in upwelling areas [*5*], thereby compensating for the losses from surface waters. Another part of the POP that sinks to depth [*3*] reaches the ocean floor, where it is buried in marine sediments. In soils and marine sediments, some phosphorus of biological origin combines with inorganic or organic materials to form very stable compounds, making this phosphorus unavailable to living organisms and thus moving it from the short to the long P cycle.

The long P cycle has a duration of hundreds of millions of years. The sediments that accumulate on the seafloor slowly form rocks [*6*], which are later transported onto continents by the tectonic plate movements (orogenesis) [*7*]. The physical and chemical weathering of these rocks, which is very slow, produces PO_4^{3-} [*8*], most of which is precipitated as chemical compounds that living organisms cannot use. Under natural conditions, the long-term weathering of phosphate rocks compensates for the losses from the short cycle.

Human societies have consumed much phosphorus during the last several decades. Phosphorus is used in particular in fertilizers, which were once in the form of naturally occurring guano (section 4.5.2) and are today in the form of superphosphate obtained by the chemical transformation of apatite. Phosphorus is also used to produce matches, and as organophosphate to produce pesticides. In addition, P is used in detergents, but concentrations of P are now regulated in many countries to reduce eutrophication of freshwaters. Most of the industrial phosphorus comes from apatite mines, which are under exploitation in small number of countries.

Human interventions in the phosphorus cycle accelerate this cycle in two major ways. The first is associated with the short P cycle, where human activity accelerates biological production [*9*], which is precisely what is intended when the soil is enriched with fertilizer, making it more fertile than it would naturally be (however, fertilizers are often used today simply to compensate for the nutrient depletion of soils). Yet, an acceleration in biological production becomes harmful in aquatic environments when is it due to eutrophication [*10*] (section 4.5.2). This occurred on large scales in freshwaters in industrial countries during the 1960–1970s, mainly due to the mass use of detergents containing PO_4^{3-}; this eutrophication has been curbed by national regulations reducing the use of phosphate in detergents and by the secondary treatment of wastewater. The second type of acceleration is associated with the long P cycle. Within the last decades, human activities have dispersed over the whole planet a chemical element, phosphorus, which had become concentrated in a few deposits by biogeochemical flows over millions of years. Some researchers have calculated that, considering the current rate of phosphorus use, the few exploitable P reserves may be exhausted in less than a hundred years (Table 4.3). As there is no substitute for phosphorus, human societies could not, in all likelihood, replace this element with any other, unlike hydrocarbons that humans hope to replace with different forms of carbon-free, renewable energy when the reserves are exhausted. Exploitable land deposits of P will be exhausted by human societies over the next decades, but their natural replenishment will take hundreds of millions of years. Will the next crisis faced by human societies be about phosphorus? Will countries fight over this essential and rare chemical element?

4.7. Biogeochemical equilibria and human societies: problems

The effects of the current anthropogenic disturbances of biogeochemical equilibria will differ according to the time scale considered. The following paragraphs briefly examine some relationships that exist between biogeochemical equilibria and human societies on three different time scales, i.e. millions of years (very long term), centuries to millennia (long term), and decades to centuries (human societies). Section 2.5 in Chapter 2 of [MON 14b] provides a wider perspective that includes ecosystems.

Over millions of years, some very long-term biogeochemical cycles have the potential to maintain the stationary state of different chemical elements in Earth's reservoirs. For example, the carbonate–silicate cycle (section 4.3.2, equations [4.6]–[4.9]) can regulate the concentration of atmospheric CO_2 and, therefore, the climate of Earth over millions of years. This type of very long-term regulation will likely cancel out most of the current anthropogenic disturbances, whose duration will have been very short compared with the very long-term biogeochemical cycles.

On a scale of a few centuries to millennia, the long-term biogeochemical cycles can mitigate some disturbances to the natural biogeochemical cycles by human societies. One example is the carbonate cycle (section 4.3.2, equations [4.1]–[4.5], which may contribute to regulating the concentration of atmospheric CO_2 and hence the climate on Earth over hundreds or thousands of years. However, its net effect is null over millions of years.

Over a few decades to centuries, human societies can themselves be affected by changes that they have caused and are increasingly causing to biogeochemical cycles. Examples of such feedback include conflicts between countries for access to nitrogen (guano, 19th Century; section 4.5.2), carbon (hydrocarbons, 20th and 21st Century; section 4.3.2) and perhaps phosphorus in the 21st or 22nd Century (apatite; section 4.6.2). In addition to these conflicts, the functioning of human societies will be perturbed in the short- or medium term by the depletion of deposits of many essential chemical elements or compounds that had accumulated in the lithosphere over millions of years (phosphorus, hydrocarbons, etc.) Finally, the huge injection of chemical elements or compounds into the hydrosphere, atmosphere and biosphere has already had and will continue to have global consequences. These include, amongst others, the warming of atmosphere and ocean, which leads to ocean deoxygenation (section 4.4.2), sea level increase, ocean acidification (section 4.3.2) and eutrophication of aquatic environments (sections 4.5.2 and 4.6.2).

This chapter has shown that biogeochemical cycles are not only interesting to researchers, but are also extremely important to human societies, because it is humans who modify the large natural cycles and are negatively affected by the resulting feedbacks. This stresses the need for better understanding the functioning of these cycles in order to prevent their disruptions as much as possible.

4.8. Bibliography

[AUT 09] MULTIPLE AUTHORS, "Special issue on the future of ocean biogeochemistry in a high-CO_2 world", *Oceanography*, vol. 22, no. 4, December 2009.

[FAL 00] FALKOWSKI P., SCHOLES R.J., BOYLE E., *et al.*, "The global carbon cycle: a test of our knowledge of Earth as a system", *Science*, vol. 290, no. 5490, pp. 291–296, 2000.

[GOR 91] GORHAM E., "Biogechemistry: its origins and development", *Biogeochemistry*, vol. 13, pp. 199–239, 1991.

[JAM 09] JAMBON A., THOMAS A., *Géochimie – Géodynamique et cycles*, Dunod, Paris, 2009.

[KEE 10] KEELING R.F., KÖRTZINGER A., GRUBER N., "Ocean deoxygenation in a warming world", *Annual Review of Marine Science*, vol. 2, pp. 199–229, 2010.

[MON 14b] MONACO A., PROUZET P. (eds.), Ecosystem Sustainablilty and Global Change, ISTE, and John Wiley & Sons, London, 2014.

[PID 06] PIDWIRNY M., "The carbon cycle", *Fundamentals of Physical Geography*, 2nd edition, Fundamentals eBook, 2006.

[RIE 07] RIEBESELL U., SCHULZ K.G., BELLERBY R.G.J., *et al.*, "Enhanced biological carbon consumption in a high CO_2 ocean", *Nature*, vol. 450, pp. 545–548, 2007.

[SAR 06] SARMIENTO J.L., GRUBER N., *Ocean Biogeochemical Dynamics*, Princeton University Press, Princeton, 2006.

[TOR 97] TORTELL P.D., REINFELDER J.R., MOREL F.M.M., "Active uptake of bicarbonate by diatoms", *Nature*, vol. 390, pp. 243–244, 1997.

[TOR 00] TORTELL P.D., RAU G.H., MOREL F.M.M., "Inorganic carbon acquisition in coastal Pacific phytoplankton communities", *Limnology and Oceanography*, vol. 45, pp. 1485–1500, 2000.

[VER 26] VERNARDSY W., *Biosphera (The Biosphère)*, Nauchnoe khimiko-technicheskoye izdatel'stvo (Scientific Chemico-Technical Publishing) Leningrad, 1926.

[VER 29] VERNADSKY W., *La Biosphère*, 2nd revised and enlarged edition, Félix Alcan, Paris, 1929 (Re-edited with a preface by Jean-Paul Deléage, Le Seuil, Paris, 2002).

[VER 86] VERNADSKY W., *The Biosphere*, English translation, Synergetic Press, Oracle, Arizona, 1986.

5

Ocean Acidification and its Consequences

5.1. Introduction

Before the Industrial Revolution, the flow of carbon from the ocean into the atmosphere was slightly higher than that from the atmosphere into the ocean (70.6 and 70 Gt C per year, respectively [IPP 07]). Industrialization, along with an increased use of fossil fuels, production of cement and land-use changes, have led to an additional flow of carbon of anthropic origin into the atmosphere. The average growth of emissions was 2% per year during the decade 1980–1990 and 3.1% per year since 2000 [PET 12]. It was 5.4% in 2010, a year during which the total of all emissions amounted to 10 Gt C [PET 12]. The concentration in carbon dioxide (CO_2) in the atmosphere has, therefore, increased from 280 parts per million by volume (ppmv) in 1750 to 394 ppmv in 2012.

The CO_2 dissolved in the surface of most of the world's oceans is in near equilibrium with the atmospheric CO_2. The increase in the rate of CO_2 in the atmosphere, therefore, results in an increase in the flow of carbon into the oceans and a reversal of the pre-industrial trend; the ocean has become a CO_2 sink with a net uptake of about 2 Gt C per year [IPP 07].

Chapter written by Jean-Pierre GATTUSO, Lina HANSSON and Frédéric GAZEAU.

The oceans, therefore, have a leading role in climate regulation and mitigation of global warming, and they have greatly contributed toward slowing the increase in atmospheric CO_2 in the last 250 years. Without the uptake of CO_2 by the oceans of 25 million tons per day (thus 30% of anthropic CO_2 emissions since the Industrial Revolution), the amount of CO_2 in the atmosphere would have already reached 450 ppmv [LEQ 09, SAB 04]. If future CO_2 emissions continue using a business-as-usual scenario, the concentration in CO_2 could reach 788 ppmv by 2100 [ORR 11].

5.1.1. *What is ocean acidification?*

Ocean acidification corresponds to a decrease in the pH of the ocean over long (10 or more years) periods of time, which is mainly caused by the uptake of atmospheric CO_2. It can also be caused by other processes, generally minor on a global scale, such as the atmospheric deposition of nitrogen and atmospheric sulfur [DON 07]. The delivery of nutrients and organic matter by rivers as well as biogeochemical coastal processes can decrease or increase the acidification caused by the absorption of atmospheric CO_2 [BOR 11, CAI 11].

Ocean acidification, sometimes referred to as "the other CO_2 problem" due to its common origins with global warming, is an expression that refers to a series of chemical modifications in seawater elicited by the absorption of anthropogenic CO_2 by the oceans. The dissolution of CO_2 in seawater leads to a disruption in ocean chemistry, including a decrease of the pH and of the concentration in carbonate ions (CO_3^{2-}) (see the Appendix, section 5.8, and Table 5.1). The term "ocean acidification", coined by Broecker and Clark [BRO 01] and popularized by Caldeira and Wickett [CAL 03], sometimes causes misunderstandings with non-scientists. Oceans will not become acidic (their pH will never be lower than 7), at least not in the foreseeable future and not away from CO_2 sources of hydrothermal origin. The pH is moving toward the acidic end on the pH scale. It has decreased by 0.1 units between 1750 and 1994 (8.2–8.1 on the total scale, Table 5.1 [ORR 11]), thus an increase in the concentration of protons, and therefore of acidity, of 30% due to the logarithmic nature of the pH scale. At the current rate of emissions, the pH could reach 7.7–7.8 toward the end of the century, which would correspond to a tripling of the average acidity of the ocean's surface in 250 years. The current rate of CO_2 uptake is probably without precedent over the course of the last 300 million years [HON 12].

See the Appendix, section 5.8, for more information on the chemistry of carbonates and biogeochemical processes.

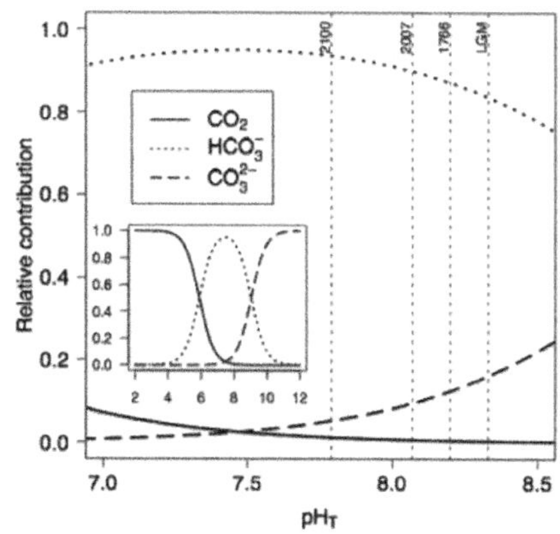

Figure 5.1. *Relative contributions of different species of dissolved inorganic carbon, gas (CO_2), bicarbonate (HCO_3^-) and carbonate (CO_3^{2-}) as a function of pH (total scale) at a temperature of 15°C and a salinity of 35. The vertical lines indicate the average pH of surface waters during the last glacial maximum (LGM) and the years 1766, 2007 and 2100 (see also Table 5.1). According to [GAT 11a]*

While the chemical perturbations are known with a high degree of certainty, the biological, ecological and biogeochemical consequences of ocean acidification only started to be studied at the end of the 1990s and remain relatively unknown to this day, despite an exponential increase of the number of studies in recent years. Initial research was dedicated to calcifying organisms, motivated by their direct need for carbonate ions during calcification. At weak pH values, seawater is under-saturated calcium carbonate ($CaCO_3$, see the Appendix, section 5.8), and is therefore corrosive to limestone, which compromises the creation of shells or skeletons and leads to the dissolution of existing structures. However, other organisms and processes can also be affected, and recent papers have provided varied and sometimes contradictory results. Several studies have adopted an approach of meta-analysis in order to summarize the results of this literature, which only continue to increase year by year [HEN 10, KRO 10, KRO 13, LIU 10].

Parameter	Unit	1766	2007	2100
Temperature	°C	18.3	18.9	21.4
Salinity	–	34.9	34.9	34.7
Concentration of phosphate	10^{-6} mol kg^{-1}	0.66	0.63	0.55
Concentration of silicate	10^{-6} mol kg^{-1}	7.35	7.35	7.35
Total alkalinity	10^{-6} mol kg^{-1}	2326	2325	2310
Partial pressure of CO_2 (seawater)	µatm	267	384	793
$[CO_2]$	10^{-6} mol kg^{-1}	9.05	12.8	24.7
$[HCO_3^-]$	10^{-6} mol kg^{-1}	1754	1865	2020
$[CO_3^{2-}]$	10^{-6} mol kg^{-1}	231	186	118
Dissolved inorganic carbon	10^{-6} mol kg^{-1}	1994	2064	2162
pH_T		8.2	8.07	7.79
$[H^+]$	10^{-9} mol kg^{-1}	6.379	8.6	16.13
Saturation in calcite	–	5.5	4.5	2.8
Saturation in aragonite	–	3.6	2.9	1.8

Table 5.1. *Carbonate chemistry of surface seawater between 1766 and 2100 [GAT 09]*

COMMENTARY ON TABLE 5.1.– Total alkalinity, partial pressure of CO_2 (pCO_2), salinity and temperature have been used for calculating all of the other parameters using the R package seacarb [LAV 10] and the dissociation constants of Leuker *et al.* [LUE 00]. The ocean and atmosphere have been considered to be in equilibrium. The temperature values, salinity, total alkalinity and phosphate in 1766, 2007 and 2100 are from Plattner *et al.* [PLA 01] using historical values of CO_2 and radiative forcing not linked to CO_2 between 1766 and 1990 and the IPCC Special Report on Emissions Scenarios (SRES) A2 scenario between 1990 and 2100 [NAK 00]. The pCO_2 values in 1766 and 2100 are from Plattner *et al.* [PLA 01] whereas the 2007 value is from

Keeling *et al.* [KEE 08]. The silicate concentration is calculated using data from Garcia *et al.* [GAR 06] between 0 and 10 m in depth and by weighing the means over the surface of each geographical cell. The pH is expressed on the total scale.

5.1.2. *A brief history of research on the acidification of oceans*

The succinct chronology of the research on the acidification of oceans by Gattuso and Hansson [GAT 11a] is summarized below.

The first studies on the pH of seawater, its spatial and vertical variation, as well as its biological consequences date back to the beginning of the 20th Century. Some of them were carried out even before the publication of the definition of pH [SOR 09]. The first studies on the biological impact of lower pH were also published around that time. For example, Moore *et al.* [MOO 06] studied the effect of acids and bases on cell growth and division of sea urchin eggs.

Some of these early studies were surprisingly advanced. McClendon [MCC 18] showed that algae tolerate a relatively large range of pH, which can vary depending on species and their life stage, and that there can be an interaction with other environmental factors. Gail [GAI 19] took interest in the combined effects of pH and temperature on the survival, reproduction and growth of a macroalgae (Fucus) and attempted to evaluate its potential for acclimation.

Powers [POW 20] suggested that the biological diversity of the Puget Sound region (west coast of the United States), as well as the average size of species such as cirripedes, is dependent on the pH. Two years later, he suggested that the "ability of fish to absorb dissolved oxygen depends on the concentration in hydrogen ions" [POW 22]. Bouxin [BOU 26] observed that the larval development of a species of sea urchin was not affected by pH changes in the range of 7.3–8.1, and was slower for pH values below 7.3 and inexistent at a pH of 6.4. Prytherch [PRY 29] showed that the pH at Milford Harbor (Connecticut, United States) varies according to the tide (7.2 at low tide as opposed to 8.2 at high tide). He observed that the failure of the hatching of oysters at low tide was linked to relatively high temperatures (higher than 20°C) and a low pH. The hatching took place at high tide at a similar temperature, but with a much higher pH (7.8 or higher). He concluded that the most important factors affecting the hatching of oysters were "the temperature of the water, the amplitude of the tide, and the concentration in hydrogen ions".

As interesting and innovative as these works were at the time, they are not particularly helpful in determining the future impact of ocean acidification. The chemistry of seawater was measured in an incomplete and approximate manner, and the pH values that were used are often extremely low. It is, however, very interesting to note that the research questions, which are currently a priority (e.g. multifactorial perturbation experiments, studies of acclimation and adaptation), had occasionally been studied a few decades ago.

We can consider that modern research began with the founding article of Revelle and Suess [REV 57], which showed that the absorption of anthropogenic CO_2 decreases the buffering capacity of seawater. Broecker and Takahashi [BRO 66] were the first to suggest that the concentration in CO_3^{2-} could control community calcification. This has been confirmed for coral reefs [SMI 74] by using mesocosms (e.g. [LAN 00, LEC 00]).

Brewer [BRE 78] gave the first direct proof that the increase in atmospheric CO_2 led to an increase of partial pressure of CO_2 (pCO_2) of Antarctic intermediate waters.

Recent changes in the carbonate chemistry have been confirmed due to oceanic time-series in the Northwest Atlantic [BAT 01], Northeast Atlantic [SAN 07], North Pacific [DOR 09] and in the Iceland Sea [OLA 09].

The first important studies using paleoceanographic tracers are those of Spivack *et al.* [SPI 93], who used the isotopic composition of boron, those of Barker and Elderfield [BAR 02] who used the size-normalized weight of the tests of foraminifera and Yu *et al.* [YU 07] who proposed the use of the boron/calcium ratio to evaluate the past changes in the chemistry of carbonates resulting from ocean acidification. Modeling has also been used to reconstruct the past changes of the chemistry of carbonates using relatively simple [ZEE 03] or more sophisticated models (e.g. [AND 05]).

Manipulation experiments in the laboratory have been carried out to evaluate the response of organisms to the changes in the chemistry of carbonates of seawater. Several of these studies have used extreme and unrealistic levels of pCO_2 or pH in the context of anthropogenic acidification. Other experiments provided little information on the parameters of the carbonate system other than pH. Nonetheless, these experiments have produced important results. For example, Swift and Taylor [SWI 66] have shown that the growth of a coccolithophore follows a bell curve relative to the pH, a similar response was later found for describing the effect of the acidification of oceans on the calcification rate of a coccolithophore [LAN 06].

The first study that aimed to study the response of an organism to ocean acidification is that of Agegian [AGE 85] who revealed a negative impact on the growth of calcareous algae. The first article on a reef-building coral highlighted a decrease of calcification as a function of decreasing $CaCO_3$ saturation [GAT 98]. Riebesell *et al.* [RIE 00] were the first to show the effects of ocean acidification on photosynthesis and the calcification of a phytoplankton group (coccolithophores). The results published in recent years have sometimes been conflicting. For example, some calcifying organisms do not seem to be affected by ocean acidification (for example, [IGL 08, LAN 06, RIE 09, ROD 10]. Several hypotheses have been suggested for explaining these differences [GAT 11b].

Modeling tools are essential for predicting forthcoming changes and the biogeochemical impacts of ocean acidification. Caldeira and Wickett [CAL 03] published a key article explaining the chronology and magnitude of the past and future changes in oceanic pH. Models have been used for estimating changes in the carbonate system in the recent past and in the 21st Century (for example, [ORR 05]), identifying areas of risk [STE 09], estimating future changes in calcification [KLE 99] as well as for investigating the feedback on the carbon cycle (e.g. [HEI 04]).

Modern research on the acidification of oceans is, therefore, very young. A bibliographical analysis of the 1,514 articles published between 1906 and 2012 highlights the exponential change of this research area over the last few years (Figure 5.2).

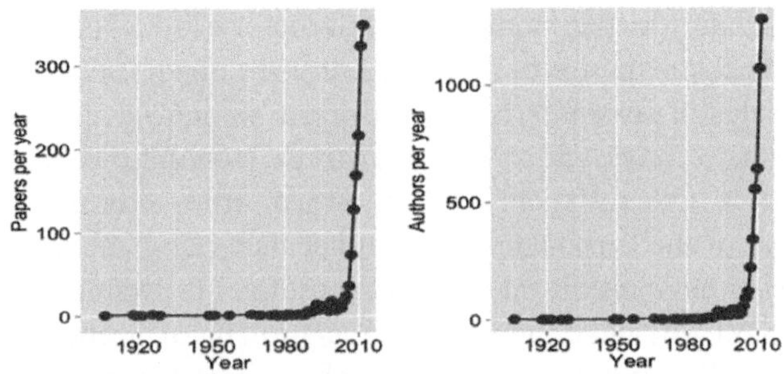

Figure 5.2. *Ocean acidification: number of publications and authors per year. See [GAT 11a] for additional information on the bibliographic database used*

The vast majority (88%) of articles were published after 2003. The number of articles published per year has increased from 22 articles in 2004 to 353 articles in 2012, thus an increase by a factor of 16.

The trend of the type of study (63% of the articles published until 2010 focused on the biological response to acidification [GAT 11a]) and on the organism and the process investigated show that the research effort has mostly been restricted to a few taxonomical groups (phytoplankton, followed by corals, molluscs and macro-algae) and processes (primary production, growth and calcification). The most studied regions are the Atlantic and the North Pacific, and the majority of studies have been carried out in the laboratory (67% vs. 33% of field studies). The research effort is, therefore, very imbalanced, with few studies dedicated to the polar regions, where the effects of ocean acidification are projected to occur sooner than elsewhere, and few studies on communities and species of economically important taxonomic groups such as fish and crustaceans. However, some of these shortcomings have attracted growing attention in the last few years. Finally, Europe has been the main contributor of articles on ocean acidification during the period 1906–2012 (41%, based on the affiliation of the first author), followed by the United States (27%).

5.1.3. *Main research programs*

Launched in May 2008, the *European Project on OCean Acidification* (EPOCA), coordinated by CNRS-INSU, was the first multinational research project on ocean acidification on a large scale (32 partners from 10 European countries). Between its launch and end in 2012, multiple national programs had been launched (BIOACID in Germany, UKOA in the United Kingdom, less coordinated efforts in China, Korea, Australia and New Zealand, Brazil and Chile, etc.) as well as a second European project dedicated to the Mediterranean (MedSeA). A US research effort is being developed and should be launched in the coming years. A new call for proposals by the European Commission was published in 2012, soliciting project propositions where acidification is studied on a larger context, including global warming and deoxygenation.

In response to the scientific community and the users of research as well as the growing concerns of Member States of the International Atomic Energy Agency (IAEA), an "Ocean Acidification International Coordination Center (OA-ICC) was launched in 2012. It is a project of the "Peaceful Uses Initiative" of the IAEA, which has the goal of facilitating and promoting activities, such as the development of an international observation network, the training of young researchers, student exchanges, the creation of databases, etc.

5.2. Observations

Seawater carbonate chemistry has undergone large changes on a geological scale as well as since the beginning of the Industrial Revolution. This section aims to characterize not only the magnitude of these changes, but also the rate at which they occurred.

5.2.1. *Past changes*

One important note must be made before considering the question of ancient episodes of ocean acidification: two parameters are necessary for characterizing the carbonate chemistry. Overlooking this requirement has led to misunderstandings in the literature. For example, the future concentration of CO_2 in the atmosphere could reach a value close to that of the Cretaceous period (145–65 Myr ago), about 2,000 ppmv. However, the Cretaceous period is a geological period during which calcification was very intense, as proven by the limestone cliffs of the Pas-de-Calais and Dover, which were formed at that time. It would, however, be a serious mistake to conclude that calcification will not be affected in a high-CO_2 future because the carbonate system will be very different from that of the Cretaceous period. Seawater with a concentration of dissolved inorganic carbon (C_T) of $2,400 \times 10^{-6}$ mol kg^{-1} in equilibrium with an atmosphere with a concentration of CO_2 of 2,000 ppmv exhibits a calcite saturation state of 1.1 (at a temperature of 15°C and a salinity of 35). However, the C_T of surface waters in the Cretaceous period were higher ($4,900 \times 10^{-6}$ mol kg^{-1}), and the calcite saturation state was 4.5 (at the same temperature and salinity). This example, taken from Zeebe and Ridgwell [ZEE 11], shows that despite their high CO_2 content, the surface waters of the Cretaceous period were highly favorable to calcification due to their high saturation in $CaCO_3$, which will not be the case in the future ocean.

The main reason for which the pH and the saturation of $CaCO_3$ are coupled today and in the near future, as opposed to the Cretaceous period where they were decoupled, is due to $CaCO_3$ compensation. The depth of the limit between waters that are saturated and under-saturated in $CaCO_3$ (saturation horizon (SH)) is subject to large vertical changes such that the supply of CO_3^{2-} and of calcium (Ca^{2+}) by rivers are, at equilibrium, equal to the sedimentation of $CaCO_3$. If $CaCO_3$ deposition is higher than its supply, then the SH moves upward, exposing a larger area of sediment to corrosive waters and reducing $CaCO_3$ burial. In the opposite case, if $CaCO_3$ deposition is lower than its supply, the SH moves downward, exposing a smaller area of sediment to corrosive waters and increasing $CaCO_3$ burial in the sediment. This compensation mechanism is very efficient over timescales longer than 10,000 years [ZEE 11] but does not operate in the current period of acidification, which is much too fast.

Several episodes of ocean acidification have been identified at a geological scale (Figure 5.3, [ZEE 11]). The major ones are the Paleocene-Eocene Thermal Maximum (PETM, ~56 Myr ago), the Cretaceous-Tertiary transition (~65 Myr), multiple oceanic anoxic events (OAE) and the Permo-Triassic (P-T, ~252 Myr). The PETM and the P-T are considered to be among the closest to current episode [HON 12] in terms of the magnitude of the change in pH, but what distinguishes the current event is its very high rate. The P-T corresponds to the biggest mass extinction of the Phanerozoic Eon, following a release of CO_2 of 13,000–43,000 Gt C in 20–400 ka, thus an annual release of carbon of 0.1–1 Gt C [HON 12] compared to 10 Gt C in 2010 [PET 12]. The current and future changes are without precedent in the last 300 million years.

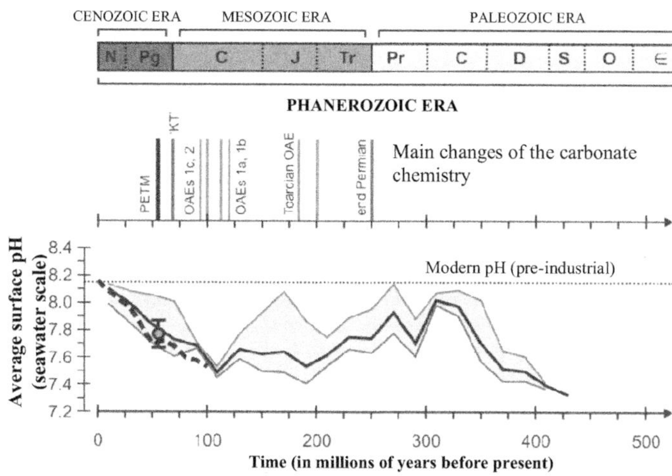

Figure 5.3. *Geological context of the acidification of oceans (according to [ZEE 11]): the major periods of acidification and the corresponding pH of surface waters on the "seawater scale" (see color section)*

5.2.2. *Recent changes*

Direct measurements of the carbonate chemistry confirm without any ambiguity that ocean acidification is a real phenomenon and that it is in progress. Orr [ORR 11] has reported on the available data. Most of the data were obtained in three long-term series that have been maintained, some for up to three decades, in subtropical locations: the ALOHA station of the *Hawaiian Ocean Time Series* (HOT) program in the North Pacific Ocean as well as the BATS (*Bermuda Atlantic Time-Series Station*) and ESTOC (*European Time Series in the Canary Islands*) stations in the North Atlantic Ocean. The decrease of pH in surface waters varies between –0.0017 and –0.0019 units per year (Figure 5.4) and does not show any significant difference between stations. The concentration in CO_3^{2-} also decreases significantly in the three stations; it is 80% higher at ESTOC than at HOT and BATS. Subsurface waters are also affected. For example, at the ALOHA station, acidification is more significant at 250 m than at the surface. It is significant up to a depth of at least 600 m at ALOHA [DOR 09] and at least 1000 m at ESTOC [GON 10].

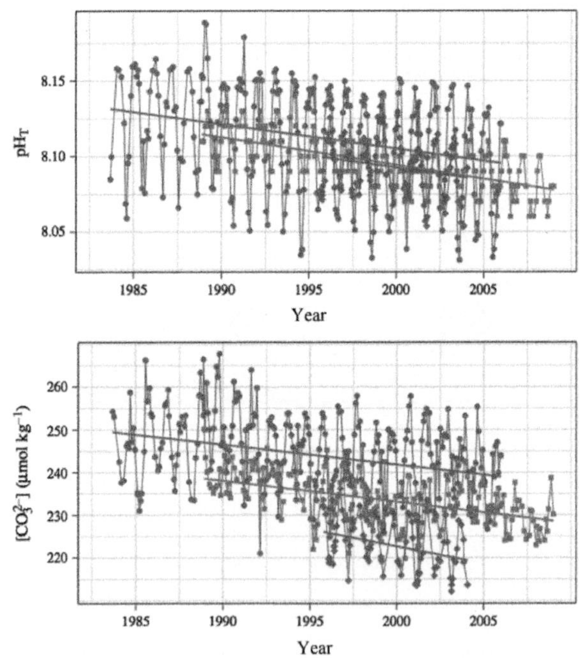

Figure 5.4. *pH (expressed on the total scale) and concentration in carbonate at the surface of the ALOHA (in green; [DOR 09]), BATS (in red; [BAT 07]) and ESTOC stations (in blue; [GON 10]) (according to [ORR 11]) (see color section)*

Other data are available with a lower measuring frequency or over shorter periods than in the three stations mentioned above. We should also mention the data collected in the Iceland Sea by Olafsson *et al.* [OLA 09], where the decrease of pH in surface waters (–0.0024 units per year) is much faster than in subtropical regions. The pH decrease below 1,500 m is 25% of that of surface waters. The depth of the saturation horizon for aragonite (one of the main forms of $CaCO_3$; see the Appendix, section 5.8) was 1,710 m in 2009. It increased at a speed of 4 m per year between 1995 and 2005, exposing an additional 800 km^2 of sea bottom each year to water corrosive to aragonite.

5.3. Projections

Attempting to predict the future consequences of the acidification of oceans requires projections of the future evolution of the carbonate chemistry. This section, mostly based on Joos *et al.* [JOO 11], summarizes the most recent estimates. Projections are carried out using models based on scenarios, which take into consideration the possible changes in the demography, economy, technology and environment. Among the scenarios used are SRES scenarios of the Intergovernmental Panel on Climate Change (IPCC; [NAK 00]) and the more recent *Representative Concentration Pathways* scenarios (Representative Concentration Pathways (RCP); [VAN 11]) which correspond to a set radiative forcing (RF) at the end of the century. There are four RCP scenarios: two reference scenarios without mitigation (RF 6 and 8.5 W m^{-2} at the end of the century) and two scenarios with mitigation (RF 2.6 and 4.5 W m^{-2} at the end of the century). The SRES and RCP scenarios have been used in different models in order to obtain projections of the climate and the carbonate chemistry.

Figure 5.5 shows the annual and cumulative carbon releases of the main scenarios. The highest emissions lead to high RF, especially when mitigation is absent or weak. When there is no mitigation, the release of greenhouse gases ranges between 70 and 250% of that of the year 2000. The increase in emissions slows down in the second half of the century due to a stabilization of the population and technological advances. The scenarios that include mitigation show a peak in emissions between 2020 and 2040 with a maximum emission that is 50% higher than in 2000 [JOO 11].

The changes in argonite saturation (Ω_a) and pH in the surface waters follow the changes in CO_2 emissions. The mean value of Ω_a, which was approximately 3.7 in the pre-industrial period, decreases to reach 1.8–2.3 in 2100 with the reference scenarios and 2.4–3.1 with the scenarios that include mitigation. pH_T decreases from a pre-industrial value of 8.18 to 7.73–7.88 and 7.90–8.05, respectively, with the baseline scenarios and with those that include mitigation. The lowest pH_T value found with the most extreme scenario is 7.6. It is important to note that the uncertainty over the projections of Ω_a and the pH is almost exclusively due to the uncertainty of CO_2 emissions.

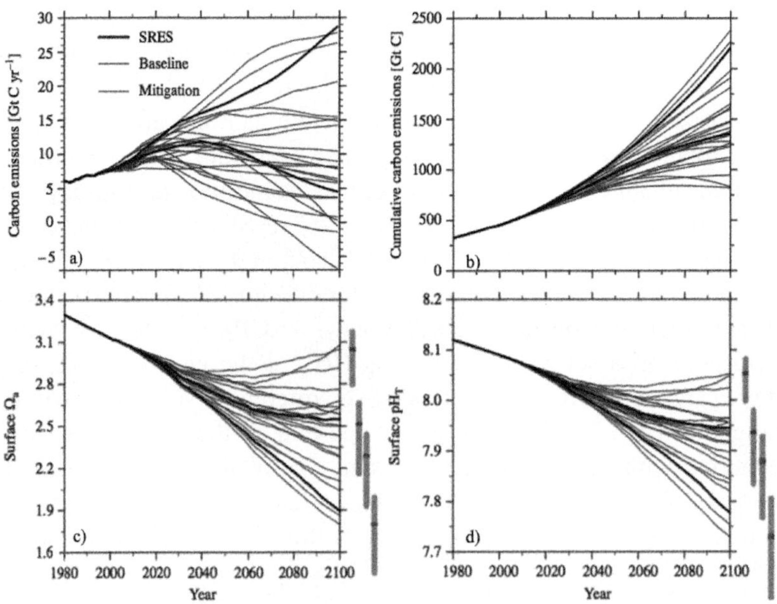

Figure 5.5. *Annual (a) and cumulative (b) CO_2 emissions used in the Bern2.5CC model and the resulting saturation in aragonite (c) and pH_T (d). The base scenarios are indicated in red, whereas the mitigation scenarios are shown in blue. The scenarios with high (SRES A2) and weak (SRES B1) CO_2 emissions are in black. The gray line shows the uncertainties in 2100 for the RCP scenarios. According to [JOO 11] (see color section)*

If the Ω_a and pH greatly diminish over the course of this century in the reference scenarios, this decrease is much less pronounced, and even turns into an increase when mitigation is implemented. This indicates without any ambiguity that mitigation is a very efficient way to limit ocean acidification.

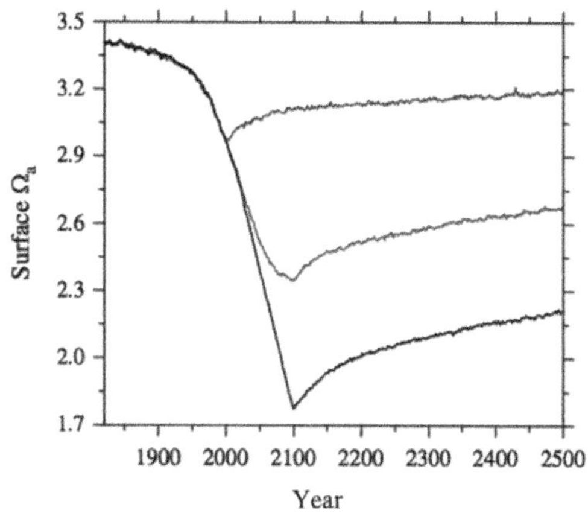

Figure 5.6. *Long-term impact of CO_2 emissions of the 21st Century on the saturation in aragonite estimated by the NCAR CSL1.4-carbon model. The three scenarios used are the following: no CO_2 emission from 2000 (blue), high emissions according to SRES A2 until 2100 and no emission afterward (red) and moderate emissions according to SRES B1 until 2100 and no emission afterward (black). According to [JOO 11] (see color section)*

Another important piece of information provided by models involves the inertia of the carbon cycle and its consequences on the long-term evolution of the carbonate chemistry. Simulations in which the emissions of carbon and other greenhouse gases are abruptly brought down to 0 in 2000 or 2100 allow us to investigate the commitment of past releases (Figure 5.6, [JOO 11]). These scenarios are not realistic but allow us to evaluate the persistence over time of the consequences of past emissions. In the first scenario, there is no CO_2 release beyond the year 2000. In the second scenario, emissions are high (SRES A2) until 2100 and are nil beyond 2100. Finally, in the third scenario, emissions are moderate (SRES B1) until 2100 and nil beyond that year. The first outcome of this study confirms that the average value of Ω_a in surface waters varies proportionally to CO_2

emissions. It decreases as emissions continue and increases as soon as emissions come down to 0. Regulation of human activity is, therefore, a powerful tool for controlling the future evolution of the carbonate chemistry. Another important finding is that the perturbation of seawater chemistry persists for several centuries. If the emissions had been interrupted in 2000, the Ω_a would not have returned to its pre-industrial level by 2500. The situation is worse when emissions are interrupted in 2100, even more so when the emissions prior to this interruption are high.

5.4. Impacts of ocean acidification

5.4.1. *Impacts on organisms and communities*

5.4.1.1. *Pelagic ecosystems*

5.4.1.1.1. Primary production

Oceanic primary production represents half of the primary production of the planet [FIE 98]. Benthic and pelagic photosynthetic organisms play important biogeochemical and ecological roles by providing more than 99% of the organic matter used in marine food webs. It was shown years ago that the increase in atmospheric CO_2 generally stimulates the photosynthesis of terrestrial plants, which is often limited by the amount of CO_2 available. In numerous marine plants, the activity of the ribulose-1,5-diphosphate carboxylase/oxygenase (RUBISCO), the enzyme responsible for CO_2 fixation, is not at its maximum at the CO_2 levels normally found in seawater. An increase in the concentration of CO_2 could, therefore, stimulate primary production.

Concerning planktonic microalgae, it seems that the effect of ocean acidification is positive, although limited, in most investigated species [RIE 11]. This probably arises from the fact that numerous species have developed, over the course of their evolutionary history, several mechanisms to overcome CO_2 limitation by increasing the concentration of CO_2 at the site of carboxylation by RUBISCO [REI 11]. These mechanisms include transmembrane pumps and/or channels capable of concentrating carbon as bicarbonate, the most abundant form of inorganic carbon in seawater, then transforming this bicarbonate into CO_2 using the enzyme carbonic anhydrase. The CO_2 concentrating mechanisms (CCMs) generally have an energy cost, and an increase in the concentration of CO_2 in seawater could be an energy gain for cells, and therefore beneficial for growth. Positive effects of acidification have been observed for several species of diatoms (e.g. [WU 10]) and dinoflagellates (e.g. [ROS 06]). The results are not clear-cut for coccolithophores for which the response of photosynthesis is either non-existent, or slightly positive (e.g. [GAO 09, MUL 10]). However, research carried out on the diazotroph (nitrogen gas fixing) algae *Trichodesmium sp.* shows an increase in the growth of these algae at elevated pCO_2 levels (e.g. [HUT 07]).

How can such diverse responses to ocean acidification be explained? Two, non-mutually exclusive hypotheses have been proposed. On the one hand, the efficiency of CCMs differs from one species to another, and on the other hand, the affinity of the RUBISCO for the CO_2 is also variable between species.

Finally, we must highlight the fact that an increase in phytoplanktonic primary production would not have only positive effects. For example, the frequency and intensity of harmful algal blooms has increased dramatically in recent years in response to eutrophication [HAL 93]. The effects of an enrichment in CO_2 on the dynamics and toxicity of these species remain relatively unknown [FU 10].

The fact that certain species seem to benefit from elevated CO_2 whereas others do not or could change the composition of communities, with consequences on primary production, the nutritional value of organic matter for higher trophic levels and the export of organic matter to the deep ocean (biological pump). Unfortunately, few studies have been carried out on planktonic communities. Some have highlighted, at low pH values, an increase in the relative abundance of diatoms (particularly large ones) at the expense of other species such as haptophytes and nanoflagellates [TOR 08]. Other works do not show an increase in the relative abundance of diatoms [FEN 09, HAR 07]. Perturbation experiments carried out in mesocosms have shown a greater consumption of dissolved inorganic carbon when the pH decreases, indicating a stimulation of the net community primary production. In some of these experiments, it has been observed that the consumption of nutrients remained the same and the C:N ratio increased, exceeding the Redfield ratio (106:16; [RED 34]). This change of the stoichiometry of organic matter could reduce its nutritional value and affect the efficiency of bacterial degradation and secondary production [RIE 11].

5.4.1.1.2. Nitrogen fixation

Nitrogen is a limiting nutrient for photosynthesis in many oceanic provinces. A larger supply of nitrogen via an increase in fixation of nitrogen gas could have major consequences on primary production. Growth and nitrogen fixation of *Trichodesmium sp.* are stimulated at elevated pCO_2 (e.g. [HUT 07]). However this species is not the only cyanobacteria performing nitrogen fixation. Only three studies focus on other species, including one on unicellular cyanobacteria *Crocosphaera* [FU 08], for which N_2 fixation is also stimulated at a high level of elevated pCO_2, and the heterocystous cyanobacteria *Nodularia spumigena* for which the two studies carried out provided conflicting results [CZE 09, WAN 12].

5.4.1.1.3. Calcification

Calcification, the formation of calcareous skeletons and shells by organisms such as calcareous algae, corals, mollusks, foraminifera, echinoderms, crustaceans and bryozoans, can be negatively affected by the decreased availability of carbonate ions associated with the increase in acidity (Table 5.1). While the primary production of coccolithophores can, in certain cases, be slightly stimulated by ocean acidification (see section 5.4.1.1.1), their calcification can be altered. Coccolithophores are the first planktonic organisms for which the impact of the elevation in CO_2 on calcification has been studied [RIE 00]. Their response varies greatly depending on the species studied, or even for different strains of a same species. In many cases, while calcification decreases with the increase of pCO_2 (for example, [RIE 00]), certain species have a bell-shaped response curve with a maximum value at the current levels of pCO_2 and a decrease on either side of this maximum [LAN 06]. Other species are particularly resistant to changes in pCO_2, and their rate of calcification does not vary when pCO_2 increases (e.g. [LAN 06]). Finally, two papers suggest a stimulating effect of ocean acidification on calcification by coccolithophore [IGL 08, SHI 09]. However, these results have been seriously questioned [RIE 11].

Given the experimental results available, it seems difficult, or even impossible, to predict the effect of the acidification of oceans on calcifying planktonic species. A field study has recently provided very interesting pieces of information on this question. Beaufort *et al.* [BEA 11] have collected coccoliths and seawater in several regions in order to document the variations in the weight of coccoliths over the last 40,000 years (Figure 5.7). They concluded that coccolithophores are generally less calcified when pH and the concentration in CO_3^{2-} decrease, which suggests that they could be seriously affected in the coming decades. However, this work also highlighted a resistant strain, the coccoliths of which have a relatively high weight at relatively low pH values. It therefore seems that this strain has succeeded in naturally adapting to an environment which does not favor calcification. It remains to be seen whether the species exposed to normal pCO_2 will be able to adapt to the rapid decrease of pH in the surface ocean projected for the coming decades.

Other planktonic organisms, such as foraminifera and pteropods, contribute significantly to the global production of calcium carbonate. Planktonic foraminifera, which have a calcitic shell, can represent up to 50% of the $CaCO_3$ present in sediments. They are, therefore, instrumental in the carbonate counter-pump [SCH 02]. The effect of changes in the carbonate chemistry on the physiology and ecology of planktonic foraminifera has not been the subject of active research until recently. Nevertheless, a positive correlation between the weight of the shell and the concentration in CO_3^{2-} has been reported in sediment samples covering multiple glacial–interglacial cycles [BAR 02]. Similar observations have been made for much more recent periods. The shell weight in recent sediments is smaller than in Holocene sediments [MOY 09]. These observations are in agreement with laboratory experiments showing a relation between the availability of CO_3^{2-} and shell weight (e.g. [LOM 10]).

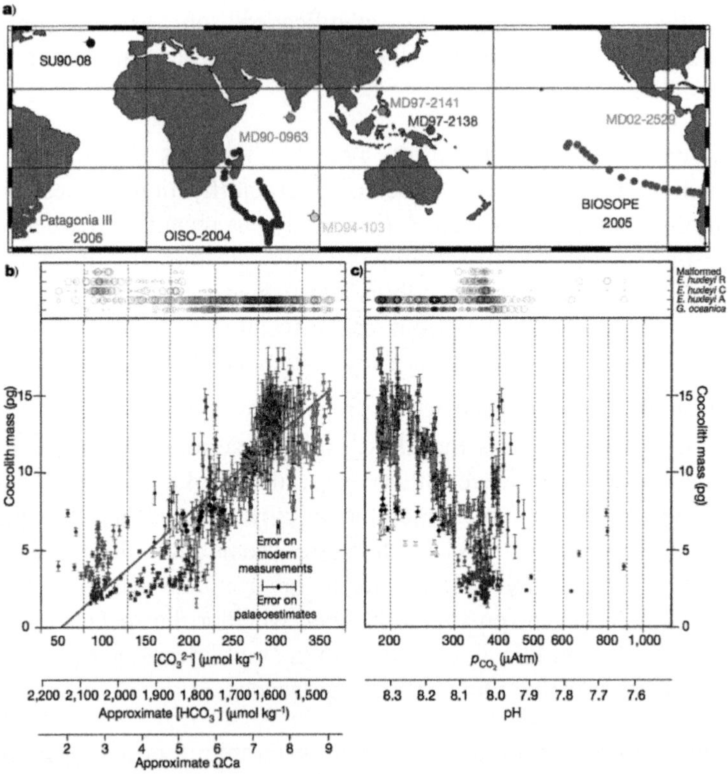

Figure 5.7. *Relationship between the mass of coccoliths from a large collection of samples covering different oceans: a) map of the locations of the samples and different timescales, and the carbonate chemistry; b) concentration in carbonate ions; c) partial pressure in CO_2. According to [BEA 11] (see color section)*

Pteropods are planktonic gastropod mollusks, the shell of which is made of aragonite, one of the most soluble forms of calcium carbonate (see the Appendix, section 5.8). These organisms are present in all oceans, but are particularly abundant in high latitude regions where they are an important component of the food chain. This is one of the reasons for which the vast majority of studies carried out on the effect of ocean acidification on pteropods that were carried out on high latitudes species (e.g. [COM 10b]) (Figure 5.8). All results indicate a significant reduction in their calcification at high pCO_2 levels, and Bednaršek *et al.* [BED 12] have shown that the shells of pteropods collected in the Southern Ocean at a depth of 200 m, where seawater is under-saturated in aragonite, show undeniable signs of dissolution. Only two studies have focused on temperate pteropods [COM 10a, COM 12]).

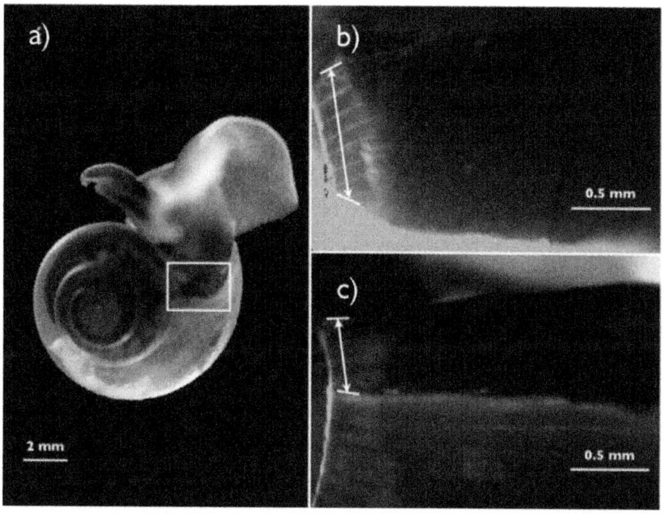

Figure 5.8. *Arctic pteropod (Limacina helicina): a) marked with the a fluorescent dye (calcein) and b) maintained at a controlled pH (pH_T ~8,09); c) at a pH projected for the end of the century (pH_T ~7,8). The arrows indicate the growth over five days after the marking of the operculum (white rectangle)*

Foraminifera and pteropods are the zooplanktonic organisms which have been most studied. There is very little information on non-calcifying zooplankton such as copepods. Some of the experiments were performed in the laboratory, but at acidity levels much higher than those projected for the next 100–300 years (e.g. [KUR 04]). Results obtained in mesocosms at pH values, which are representative of the values expected during the course of this century, suggest that the acidification of seawater does not have any effect on the structure of the mesozooplanktonic community [NIE 13], the grazing ability of microzooplankton [SUF 08] and the filtering and production of copepod eggs [CAR 07]. Significant effects on the larval development of copepods [CAR 07, RIE 08] and cirripedes [NIE 13] have nevertheless been reported.

5.4.1.1.4. Viruses and bacteria

Only a few studies have investigated the effects of ocean acidification on viral and microbial processes and diversity. These microorganisms play a major role in biogeochemical cycles through their capacity to remineralize particulate and dissolved organic matter and to regenerate nutrients used by primary producers [AZA 83]. In the framework of mesocosm experiments, changes in the bacterial community have been observed in free bacteria, but not in bacteria attached to particles [ALL 08]. These changes in bacterial diversity had no impact on total bacterial production. Modifications of bacterial production have also been observed in another mesocosm experiment [GRO 06], together with positive effects on the activity of some enzymes. Several subsequent studies have since highlighted the varied effects of elevated pCO_2 on bacterial enzymatic activity [PIO 10]. For more details on these aspects, we advise readers to consult the meta-analysis of Liu *et al.* [LIU 10] and the summary chapter of Weinbauer *et al.* [WEI 11].

5.4.1.1.5. Fish

It has long been considered that the acidification of the ocean does not have any significant impact on fish. Historically, studies of hypercapnia (elevated pCO_2 level) in fish have focused on acid–base balance and have generally been carried out at pCO_2 levels of 10,000–50,000 μatm [ISH 04]. However, several recent studies have shown that the ocean acidification can affect fish at CO_2 levels representative of those expected in the coming decades. Among noteworthy observations is the influence of ocean acidification on the formation of otoliths of the white bass [CHE 09], the survival and growth of the North American silverside [BAU 11] and on the respiratory physiology and acid–base balance of the gulf toadfish [ESB 12]. Elevated CO_2 also seem to cause irreparable damage to the tissues of cod larvae [FRO 12]. Other much less expected effects have been reported for reef fish. Acidification affects the olfactory ability of certain species, disrupting habitat selection [MUN 09], hearing abilities [SIM 11] and behavior toward predators (Figure 5.9 [MUN 10]). These effects are associated with the inhibition of certain neurotransmitters [NIL 12] and could have a negative impact on reef fish populations and their recruitment [MUN 10].

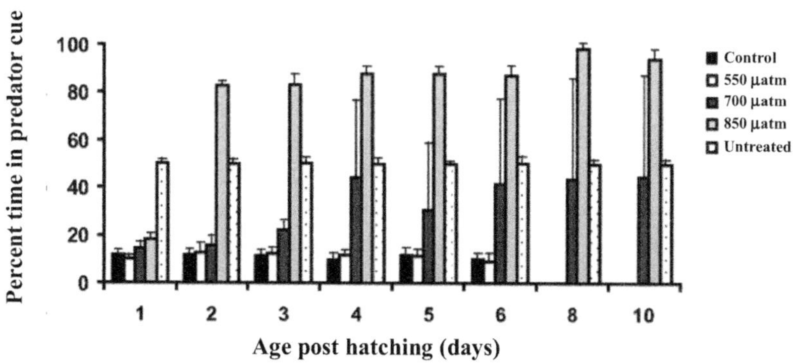

Figure 5.9. *Effects of elevated CO_2 concentrations on olfactory ability of clownfish larvae. Clownfish were exposed to 390 (control), 550, 700 or 850 ppm CO_2 from hatching, and their behavioral responses were assessed at regular ontogenetic stages using a flume chamber where they could freely move between a stream of water containing the predator cue and a stream of water without the cue. Shown is mean time (±SD) that clownfish larvae spent in the stream of water containing the predator odor. The untreated category shows the time larvae spent on one side of the chamber when neither water stream contained the predator cue. according to [MUN 10]*

5.4.1.2. *Benthic ecosystems*

5.4.1.2.1. Macroalgae and seagrasses

The positive effect of the enrichment in CO_2 on photosynthesis has been observed in certain macroalgae (e.g. [GAO 93]) and marine plants (e.g. [PAL 07]). The phanerogam *Posidonia oceanica*, an iconic plant of the Mediterranean that is lacking an efficient carbon concentration system, seems to benefit from the natural increase in CO_2 next to CO_2 vents, in contrast to calcareous algae and animals (Figure 5.10, [HAL 08]), including those living on the fronds of Posidonia (calcareous epiphytes, [MAR 08]). A recent study has highlighted a negative effect of decreased pH on a phanerogam: a large decrease in phenolic compounds as a function of increasing pCO_2 [ARN 12]. Yet, these substances protect leaves from grazing, which suggests that certain marine plants could be much more sensitive to grazing by fish and sea urchins despite the stimulation of their photosynthesis.

5.4.1.2.2. Corals and calcareous algae

In contrast to plants and fleshy algae, the response of the primary production of corals and calcareous algae, which are the first organisms to have been studied in the context of ocean acidification [AGE 85, GAT 98], is non-existent or limited, while the effect on calcification is negative for the vast majority of species studied [AND 11]. Furthermore, the simultaneous increase of pCO_2 and temperature generates a larger decrease in calcification than increase in pCO_2 alone [ANT 08, REY 03]. However, the effect of elevated CO_2 on calcification is generally less pronounced or non-existent in temperate corals and in deep-sea corals that exhibit slower growth [FOR 12, MAI 12, MAI 13, ROD 10]. Net calcification of the zooxanthellate Mediterranean coral decreases significantly with the decrease of the pH [ROD 11], and the effect of acidification is clearly higher when pH decreases concomitantly with an increase in temperature.

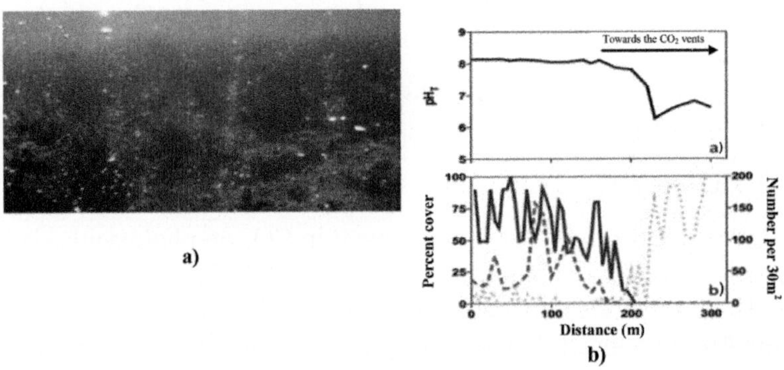

Figure 5.10. *a) Photograph of a Posidonia bed next to a CO_2 vent (Castello Aragonese, Ischia, Italy; © J. Hall-Spencer). b) Top figure shows changes in pH along a transect toward CO_2 vents; bottom figure shows along the same transect, percent cover (left axis) of calcareous (bold line) and non-calcerous algae (dots) as well as abundance (right axis) of sea urchins (dotted line). Data provided by Hall-Spencer et al. [HAL 08] (see color section)*

Tropical and temperate calcareous algae are among the most sensitive calcifying organisms to ocean acidification . This is due, at least partly, to the fact that their skeleton is made up mostly of magnesian calcite, a particularly soluble (and therefore unstable) form of CO_3^{2-} . Studies carried out until now have shown a negative effect of acidification on their recruitment, survival, growth and, unsurprisingly, their calcification rate [KUF 07, MAR 09]. It seems that algae, as with corals, are more sensitive to acidification when it is accompanied by an increase in temperature [DIA 12], which could lead to significant modifications in the structure and function of benthic communities dominated by these organisms in the coming decades. Due to their abundance, ecological importance and key contributors of carbonate deposits (and therefore habitat creation), the vulnerability of corals and calcareous algae to the increase of pCO_2 would probably have major consequences on coastal ecosystems.

5.4.1.2.3. Mollusks

Like corals and calcareous algae, benthic mollusks also provide important ecological services. They regulate the flow of energy and nutrients in coastal ecosystems, offering a habitat for numerous benthic organisms and represent an important source of food for birds. They also have a non-negligible market value and participate in the economy of many coastal regions. Shellfish aquaculture has increased by close to 8% per year over the course of the last 30 years, to create a market of more than 13.1 billion US dollars in 2008 [FAO 09]. The vast majority of mollusks have calcified external structures which sustain their tissue and protect them from predators and environmental perurbations. Depending on the species, this shell is formed of aragonite, calcite or of a mix of these two minerals.

The effect of ocean acidification on shell formation depends on the species and can even vary within strains of the same species [GAZ 07, RIE 09]. It has also been shown that their basal metabolism could be durably modified by the decrease of the pH projected for the end of the century, with potentially lethal consequences in the long term (e.g. [PAR 12, TAL 11]). While fertilization does not seem to be affected by increases in seawater acidity (e.g. [VAN 12]), embryonic and larval developments are very sensitive, with a decrease in size and survival of larvae, an increase in the number of abnormal larvae and an increased duration of the planktonic life stage (e.g. [GAZ 10]). The larval stage is a key step in the lifecycle of mollusks; small perturbations during the larval stage can compromise the sustainability of mollusk populations. The success rate of the larval stage does not exceed 2% in nature [BAY 76]; it is easy to understand that an additional perturbation, such as ocean acidification, could have disastrous consequences on these organisms. This is already the case on the coast of California where deep-sea waters rich in CO_2 rise to the surface due to a *seasonal* upwelling [FEE 08]. This natural phenomenon is reinforced by ocean acidification and results in pH levels detrimental to the development of Pacific oyster larvae cultivated in hatcheries [BAR 12]. Intense collaboration between researchers and the aquaculture industry has resulted in new management techniques including the continuous monitoring of the pH of the facilities. Seawater pumping is interrupted when the acidity of the water exceeds a critical threshold.

We should stress that mollusks have colonized, over the course of their evolution, very different environments and are even found in extreme ecosystems such as deep hydrothermal vents, which can be acidic (pH_{NBS} as low as 5.36 [TUN 09]). It is, therefore, certain that these organisms have the molecular and physiological capability to resist such hostile environments. The question is whether populations which have been exposed for thousands, or even millions, of years to relatively high pH levels will be able to adapt to the rapid increase (on the evolutionary scale) in seawater acidity. One recent study has shown that the effect of ocean acidification on the larvae of an Australian oyster is lessened when the breeders are acclimated to elevated pCO_2 levels prior to reproduction [PAR 12]. Finally, as in other taxonomic groups, interactions between changes in the carbonate chemistry and changes in other environmental parameters, such as temperature and salinity, are poorly known for mollusks.

5.4.1.2.4. Echinoderms

In contrast to mollusks, echinoderms are, in general, relatively resistant to ocean acidification. No lethal mid-term impact has been observed, but sublethal effects have been reported. For example, positive modifications have been measured in sea urchins [RIE 09] and brittle stars [WOO 08]. Other physiological alterations have been observed such as a disruption in the acid–base balance [MIL 07], a decrease of the efficiency of the immune system [HER 11], as well as an increase in respiration associated with an imbalance in the growth of tissues and calcification [WOO 08]. Echinoderms are, like mollusks, meroplanktonic organisms; in other words, they have a planktonic larval stage. Some experiments have shown a decrease in the sperm motility of sea urchin [HAV 08] and holothurian [MOR 09], suggesting a significant reduction of the fertility at low pH levels. Sea urchin larvae and juveniles respond negatively with higher rates of mortality [SHI 05]. However, these lethal effects are not always observed (e.g. [DUP 10]). Embryos appear to be rather resistant to elevated CO_2 and the first effects generally appear during the first larva stage after hatching, most of the time with a delayed growth at low pH levels (e.g. [MAR 11]).

5.4.1.3. *Unknowns and experimental approaches*

Our understanding of the consequences of ocean acidification on marine biology is restricted by the limited information available on the response of communities and ecosystems over timescales representative of the lifecycles of organisms and on the interaction with other environmental changes, such as global warming. Most of the data was collected during perturbation experiments performed on species isolated from their natural environment, often in the laboratory. It is not certain that the results of such experiments can be extrapolated to natural systems due to the lack of information on the impact of ocean acidification on ecological interactions, such as predation, competition and mutualism.

Several approaches can be used to address this shortcoming. Natural CO_2 gradients can provide valuable information. This is the case, for example, of estuaries where pH gradients often exist between the river and river mouth. Discriminating the effects of the changes in pH from those of other environmental parameters, such as salinity or the concentration in oxygen, can be particularly arduous. Observations carried out on greater spatial scales can also be considered, for example, the effects of latitudinal variation of the saturation in $CaCO_3$ on calcareous shells. Changes in weight, thickness and the makeup of the shells of tropical and polar invertebrates have been attributed to the decrease in the saturation in $CaCO_3$ at higher latitudes [WAT 12]. Even in this case, it is difficult to distinguish the effects of acidification from those of other parameters such as temperature and salinity, which can also exhibit pronounced latitudinal changes. Well-known examples of environments with a strong pCO_2 gradient are zones where volcanic activity releases CO_2. CO_2 is sometimes the major component in the emitted gas, generating a large pH gradient away from the CO_2 vents. This is the case, using the example of a site located in the Gulf of Naples (Italy), where the composition of benthic communities varies across the pCo_2 gradient, with calcifying organisms progressively disappearing when we approach the site where the gas bubbles are released and where the marine phanerogam *Posidonia oceanica* benefits from the high availability of CO_2 (Figure 5.10, [HAL 08]). These environments are naturally acidified, which allow us to take into account the interactions between organisms over long periods of time. They, therefore, provide crucial information on the effects of the acidification of oceans on communities and on the potential for adaptation of other organisms. However, spatial and temporal changes in pH are sometimes very pronounced, which complicate the interpretation of the data and the extrapolation of results to other sites. Finally, CO_2 vents are interesting for studying organisms and benthic communities; however, it is not of much interest for planktonic organisms and communities.

With regard to plankton, multiple perturbation experiments at a community level have been carried out in recent years, with some on large-sized mesocosms [ENG 05, RIE 13] (Figure 5.11).

Figure 5.11. *Planktonic mesocosms used in the context of the European project MedSeA (Mediterranean Sea Acidification in a changing climate). a) Aerial view of a group of three mesocosms (© A. Sallon), b) underwater view of a group of three mesocosms (© A. Sallon), c) in situ benthic enclosures (Coral-Proto Free Ocean Carbon Enrichment System; CP-FOCE; © University of Queensland) used on Heron Island (Australia; [KLI 12])*

Experiments in mesocosms were also carried out early on to study benthic communities [LAN 00]. Hale *et al.* [HAL 11] carried out a two-month experiment aiming at quantifying the effects of a reduction in pH and an increase in temperature on a natural community of marine invertebrates. It highlighted the changes in the structure of the community with a notable reduction in the diversity in response to a drop in pH. While mollusks appear very sensitive, with large decreases in abundance and diversity in response to an increase in acidity and temperature, the response of other organisms, such as annelids, is more complex with a decline in abundance with a decrease of pH and an increase in abundance with an increase in temperature. Several recent studies have looked at the effects of ocean acidification on micro- and macro-organisms living within the sediment and which are responsible for bioturbation and bioirrigation. These effects lead to modifications in the flow of nutrients between the sediment and the overlying water [WID 11]. Recently, Campbell and Fourqurean

[CAM 11] reported the development of a technique for modifying the pH in submerged coastal benthic communities by using open-top transparent chambers. Other systems have been developed, initially for deep-sea environments [KIR 09]. They have recently been adapted and used on a coral reef (Figure 5.11, [KLI 12]) and on marine plant communities [ARN 12]. Other systems are in development, which will allow, in the short term, for the studying of a significant number of marine communities and regions.

Another major obstacle for predicting the effects of future environmental changes, including ocean acidification, is the lack of knowledge on the potential for acclimation and adaptation. The vast majority of experiments carried out to date have never exceeded a few months in length. Yet, organisms such as mollusks and sea urchins can live for several years, and have lifecycles comprising planktonic larval stages and benthic adults. A very limited number of studies have involved the links between these life stages as well as the links existing between two successive generations. The study by Parker *et al.* [PAR 12] is very interesting as it is one of the few which shows that breeders are capable of bestowing a certain plasticity to their descendants, even if they have themselves been raised in suboptimal environmental conditions. It is crucial that similar studies are performed, and that the use of molecular tools (e.g. [HAM 12]) becomes more widespread.

Finally, as mentioned above, very few studies have focused on the effects of multiple simultaneous perturbations. The ocean will be subject to at least three global perturbations, all linked to the increase in greenhouse gases, such as CO_2. Like ocean acidification, the warming of seawater (0.7°C over the last 100 years and 3°C or more between now and 2100) is an unavoidable reality. The ocean's deoxygenation corresponds to the loss of dissolved oxygen resulting directly and indirectly from global warming, as well as the massive delivery of organic matter that has led to local eutrophication. By 2100, the ocean could lose 1–7% of its global supply of oxygen [KEE 10]. Unfortunately, very few studies have involved the combined effects of these environmental modifications. Yet, as we saw earlier, the effects of acidification can be exacerbated at high temperatures (e.g. [REY 03]). The lack of available data is a shortcoming which should absolutely be addressed.

5.4.2. *Impacts on biogeochemical cycles*

CO_2 exchanges between the ocean and atmosphere play a particularly important role in the carbon cycle and global climatic changes. As mentioned before, the ocean absorbs about 25% of anthropic CO_2, however, this uptake can be affected by feedback loops caused by the impact of acidification on physical and biological processes responsible for the CO_2 uptake. The ability of oceans to store anthropogenic CO_2 depends, firstly, on the "solubility pump", arising from the fact that the vast majority (~90%) of dissolved inorganic carbon is in the form of bicarbonate (see the Appendix, section 5.8). Therefore, most of the CO_2 penetrating the ocean through the air–sea interface is very quickly transformed into bicarbonate and thus avoids degassing: this is the buffering effect. The solubility of CO_2 increases with the drop in temperature; the solubility pump is, therefore, more active at higher latitudes, where deep water is formed. This CO_2, which is "chemically" trapped at the surface, is "physically" pulled toward the bottom of the ocean, where it circulates for hundreds of years. The "biological pump" also enables the ocean to store atmospheric CO_2 through the downward export and burial in sediment of a small fraction of the organic matter produced at the surface by autotrophic organisms. In contrast, the production of $CaCO_3$, or carbonate "counter-pump", is a source of CO_2 (see equation [5.5] in the Appendix, section 5.8), which, therefore, limits the ability of oceans to store atmospheric CO_2 [FRA 94].

Ocean acidification can affect these pumps and thus regulates the concentration in CO_2 of the atmosphere and climate. The most important feedback is the one related to changes in the buffering capacity of the ocean. The ability of the ocean to store CO_2 is inversely proportional to what is commonly referred to as the Revelle factor, in other words the ratio between the relative changes in pCO_2 and dissolved inorganic carbon (C_T):

$$R = [\delta pCO_2 / pCO_2] / [\delta C_T / C_T]$$

The smaller this ratio is, the better the capacity of the ocean for absorbing CO_2. Ocean acidification increases the Revelle factor, which leads to a drop in the ability of seawater to absorb CO_2. Consequently, the oceanic sink for CO_2 should decrease in the future; this is a positive feedback that increases the atmospheric concentration in CO_2. Furthermore, the solubility of CO_2 and the chemical balance between the different types of dissolved inorganic carbon depend on the temperature and salinity (see the Appendix, section 5.8). Both will be modified in the coming decades leading to a decrease in the solubility of CO_2 [ORR 11] as well as an increase in stratification and a slowing down of the formation of deep water [SAR 98].

With regard to the biological pump, it was mentioned in the previous section that photosynthesis can benefit the increase in CO_2. It is important to note that an increase in photosynthesis does not automatically mean an increase in the biological pump. What is important in this case is the net community production, in other words the net balance between photosynthesis and the mineralization of fixed carbon (respiration). It is this organic surplus of carbon that is exported to depths, generating a net flux of atmospheric CO_2 in the ocean. As mentioned earlier, conflicting results have been obtained depending on the phytoplanktonic species considered and not much data are available for communities and ecosystems. Nitrogen-fixing cyanobacteria is the group which has the most consistent data. They could provide an increased amount of nitrogen for primary production, but current data, carried out mostly on one species, do not allow extrapolation to the world ocean. It has already been mentioned that multiple mesocosm experiments have highlighted a change in the C:N ratio of the organic matter produced [RIE 07]. The global impact of this stochiometric change has been tested by Oschiles *et al.* [OSC 08] who concluded that the resulting negative feedback will be relatively limited. The lack of data for planktonic communities makes it impossible to currently estimate the sensitivity of the biological pump to ocean acidification and does not enable us to estimate the amplitude of the possible feedback to the climate.

Given that the precipitation of $CaCO_3$ increases CO_2 partial pressure, a decrease in calcification in response to ocean acidification (see previous section) could favor CO_2 storage in the ocean. This would be a negative feedback to the increase in the atmospheric concentration in CO_2 [GEH 11]. There are few current global models taking this process into account, partly because the large variability of responses make its parametrization difficult [GEH 07]. Despite this limitation, all models converge toward a small associated reduction in pCO_2 in response to decreased calcification. Beyond the direct effect of a decline of calcification on CO_2 fluxes, reduced rates of calcification could significantly affect the export of organic matter to the deep ocean [GEH 11]. Indeed, calcium carbonate acts as a ballast that facilitates carbon export through sedimentation. A decreased production of calcium carbonate could, therefore, decrease the exportation of organic carbon toward deep waters, reduce the efficiency of the biological pump, and therefore act as positive feedback to the increase in atmospheric CO_2. Barker *et al.* [BAR 03] were the first to consider this mechanism. Their study of sensitivity shows that this positive feedback is of the same scale as the negative feedback associated with the drop in calcification.

5.4.3. *Economy and society*

The evaluation of the economic and societal impacts of ocean acidification is complex and often hampered by the difficulty in extrapolating scant and sometimes inconsistent scientific data. It has already been mentioned that there are only a small number of studies on species of commercial interest. The experiments have often been carried out in laboratories, in conditions which do not represent the natural environment well, and over short durations, leading to results sometimes contradictory or different depending on the species considered. Works on the potential for adaptation or acclimation are almost non-existent, as well as for the response of communities and ecosystems. Occasionally, the processes measured are of limited interest for economists, for example, the reduction in calcification is not particularly useable as is. An effect on the edible part of the organism would be more interesting in an economic and societal context [HIL 13]. Generally, it is difficult to associate a cost with a service offered by an ecosystem, since it often involves indirect services which are inherently difficult to estimate. Hilmi *et al.* [HIL 13] provided recommendations aiming at producing data useful in economic studies.

There are, therefore, many challenges, but the economic aspect is essential for stimulating political action against the harmful effects of acidification. By evaluating the feasibility of solutions, as well as political and environmental priorities, decision makers must understand the value services and take into account the costs associated with taking action or inaction. The lack of economic data has been identified in several documents on ocean acidification and brought to the attention of decision makers (e.g. [PRI 09]). Recently new efforts have been made. For example, the first international meetings on the economics of ocean acidification were organized in 2010 and 2012 [IAE 11, HIL 13]. This theme is increasingly more present in scientific conferences, such as the Third International Symposium on the Ocean in a High-CO_2 World$_2$. The few recent studies presenting the first economic evaluations of the effects of acidification [BRA 12, COO 09, MOO 11, NAR 12] are briefly summarized below.

5.4.3.1. *Fishing, aquaculture and food safety*

Fishing covers 15% of the protein requirements for 3 billion people, it is the main protein source for 1 billion inhabitants, and generates 38 million jobs, 95% of these in developing countries [TUR 12]. Revenues in fishing are 80–85 billion US dollars per year, and 520 million people are directly or indirectly dependent on fishing [SUM 11]. Even if acidification is far from being the only threat to this industry, these numbers show that consequences can be significant in terms of employment and food security.

The impact of ocean acidification on fishing and aquaculture can be direct, through harmful effects on the species or ecosystem of interest, or indirect, through alterations in the food web (impacts on prey or predators of species of commercial interest) or modifications of habitats.

There are few studies on the direct effects of acidification on commercial organisms, and the majority have been dedicated to calcifying organisms due to their presumed sensitivity to acidification. For example, Gazeau *et al.* [GAZ 07] showed a drop of 25% in calcification in the blue mussel and 10% in the Pacific oyster. Meta-analyses [HEN 10, KRO 10, KRO 13] have confirmed this negative effect on the calcification of bivalve mollusks, while the effects on crustaceans and cephalopods seem less pronounced. The results involving mollusks have paved the way for the first economic studies on the acidification of oceans aiming to estimate potential losses. For example, Narita *et al.* [NAR 12] calculated that the loss of production of mollusks caused by the acidification of oceans could cost more than 100 billion US dollars in 2100.

Although there are few studies available on the economic and societal effects of ocean acidification, there does exist a very tangible example. The production of oysters on the Northwest coast of the United States, with a value of 278 million US dollars in 2009, is already suffering from damage linked to upwellings of deep waters with a low pH. In 2006, several oyster hatcheries suffered from very high mortality rates. After several years of research, and due to a partnership with scientists, the Whiskey Creek Shellfish Hatchery in Oregon established a link between the mortality and pH of the seawater that was being pumped into its installations [BAR 12].

The effects of acidification on fish of commercial interest are much less known despite being the dominant product in the global capture (69 million tons per year, thus 85% of the total weight). On the whole, acidification seems to have few direct effects on the reproduction and physiology of fish, even at very high levels of CO_2 [BRA 13]. In contrast, studies carried out on tropical fish have highlighted behavioral effects. Acidification affects sensory systems involving smell, vision and hearing. This leads to behavioral changes, such as a sudden preference for an inappropriate habitat or a decline in the ability to hear predators. The economic consequences of acidification could reveal themselves to be very significant if these effects also exist in commercial species [BRA 13].

Indirect effects on fishing can also occur through changes of the food web. For example, several studies suggest that ocean acidification threatens the existence of pteropods at high latitudes. Yet, these organisms are important prey, especially for commercial species, such as pink salmon. Current data do not enable us to make any conclusions about effects on the resource since, in the hypothetical situation where pteropods would begin to disappear, it is unknown which species would occupy the ecological niche left vacant and whether these species would have the same nutritional value for pink salmon.

5.4.3.2. *Services associated with coral reefs*

Coastal ecosystems (coral reefs, mangroves, salt marshes, seagrass beds, etc.) are instrumental in the social life and economy of many regions. Coral reefs, already under threat from intense perturbations (global warming, pollution, tourism, overfishing, etc.), are among the ecosystems that are the most sensitive to ocean acidification due to its impact on calcification of the major reef-building organisms (corals and calcerous algae). According to Frieler *et al.* [FRI 13], 70% of coral reefs will be largely degraded between now and 2030 under the effects of global warming; acidification acting as an additional stress. Coral reefs have multiple interests for society and the economy. They protect coasts from erosion and flooding, they form the basis for a strong tourism industry and are diversity *hotspots*, offering a habitat, breeding grounds or acting as a source of food for numerous species, including fish of nutritional, economic and ecological interests [TUR 11]. The dependence on these services offered by coral reefs is most pronounced in some of the poorest countries.

There are very few economic studies on coral ecosystems, and the indirect nature of their services makes evaluation difficult. Coral reefs provide food, revenue and protection for close to 500 million people [HOE 11]. Their annual value has been evaluated at 30 billion US dollars [CES 03]. Brander *et al.* [BRA 12] have estimated that damages linked to the degradation of coral reefs through ocean acidification could reach 870 billion US dollars between now and the end of the century. Recent studies are very pessimistic regarding the future of these ecosystems [ANT 11, FRI 13], and the threat of their disappearance is one of the major reasons in the plea for an urgent consideration of the problem of ocean acidification by decision makers.

5.5. What are the solutions?

Ocean acidification is already under way and its effects will be felt for hundreds of years even if CO_2 releases cease today. Nevertheless, mitigation can make a big difference. Three types of solution are available [BIL 13]: reduction of CO_2 emissions, geoengineering and adaptation (Figure 5.12).

Ocean acidification is a global problem that requires a global solution. The most efficient and least costly long-term solution is the reduction of CO_2 emissions into the atmosphere. This is the number one option recommended by documents intended for decision makers [PRI 09, BIL 12, LAF 11, OCE 09, OCE 10]. The inability to find an agreement during international negotiations reflects the difficulty of implementing this recommendation.

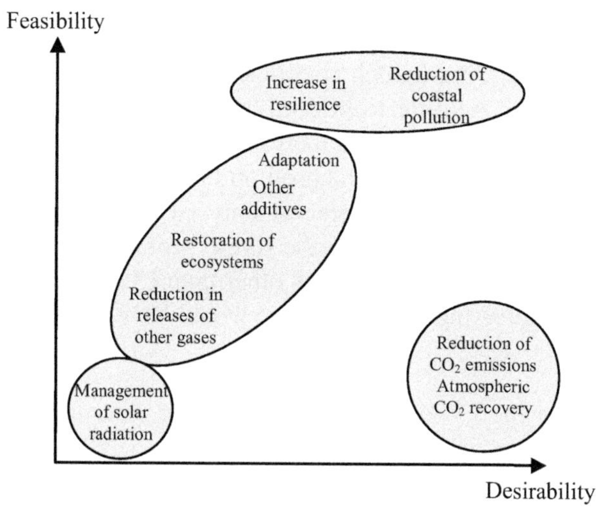

Figure 5.12. *Feasibility and desirability of the main solutions to ocean acidification (according to [BIL 13])*

5.5.1. *The reduction of CO_2 emissions*

The only international political structure mandated for reducing CO_2 emissions is that of the United Nations Framework Convention on Climate Change (UNFCCC), which seems to be an adequate framework for addressing ocean acidification [BIL 13, HAR 12, TUR 12]. The 18th session of the Conference of the Parties of the UNFCCC took place in Doha, Qatar, in November 2012. Unlike previous sessions, there was no session dedicated to the oceans. Ocean acidification and the oceans in general are not yet on the political agenda despite the significant and repeated efforts of the scientific community and research users. This situation will perhaps change in the near future. The UNFCCC is advised by IPCC. Unlike the last report of the IPCC released in 2007, ocean acidification is well covered in the fifth assessment report.

Kim [KIM 12] called in to question the use of the UNFCCC as well as that of other forums tasked with addressing ocean acidification and has proposed the creation of a new multilateral framework. His argument is based on the fact that the UNFCCC is above all tasked with addressing climate change, and is therefore not a suitable framework for addressing ocean acidification, which is not an element of climate change even though acidification and climate change share the same origin. Acidification does not formally fall within the mandate of the UNFCCC, and is therefore outside its legal framework. Furthermore, the UNFCCC is focused on the reduction of greenhouse gases in general, whereas in general only CO_2 is relevant to ocean acidification [BIL 13]. Insofar as the Kyoto protocol offers the choice of reducing greenhouse gas releases other than CO_2, it is theoretically possible that addressing climate change would have no effect on ocean acidification [HAR 12].

In order for the UNFCCC to offer adequate measures for tackling ocean acidification, it would need to:

– recognize that acidification is a problem on its own, with the same cause as climate change while remaining distinct from it;

– recognize that certain methods considered for overcoming global warming would be ineffective against ocean acidification;

– address the mitigation of CO_2 separately from that of other greenhouse gases;

– take ocean acidification into account in discussions aiming to define the acceptable level of CO_2;

– establish an ocean acidification index concurrent with the temperature.

Harrould-Kolieb and Herr [HAR 12] recognized that these are major challenges, but believe, in contrast to Kim [KIM 12], that the UNFCCC offers a better framework for preparing for the mitigation of ocean acidification and even proposed that the UNFCCC begins to exclusively focus on CO_2 emissions in order to increase the chances of simultaneously addressing climate change and ocean acidification.

5.5.2. *Geoengineering techniques*

Geoengineering is the deliberate and large-scale manipulation of the environment of the planet in order to counter anthropogenic climate change [THE 09]. There are two groups of techniques: those aiming to reduce the CO_2 that is already present in the atmosphere (carbon dioxide removal, CDR) and those aiming to reflect part of the solar radiation *and heat back into space* (solar radiation management, SRM). The potential effects of these techniques on ocean acidification have been examined by Williamson and Turley [WIL 12].

CDR techniques include the following: sequestration of the CO_2 in sub-seafloor storage, ocean fertilization through the supply of nutrients such as iron and basic products such as bicarbonate, carbonated minerals, or pulverized calcerous rocks in order to increase pH and $CaCO_3$ saturation. Some of these techniques, such as fertilization, which aims at stimulating phytoplankton growth, and therefore increasing the biological assimilation of atmospheric CO_2, only shift the problem elsewhere. A large-scale fertilization would cause a sequestration of CO_2 in deep waters following sedimentation and the decomposition of additional organic matter created at the surface. This would, therefore, increase the rate at which pH decreases in deep waters [CAO 10, OSC 10].

SRM techniques aim to counter climate change and have very little direct effect on ocean acidification. They were developed to reduce the warming of the planet, for example, by reflecting solar radiation through the injection of sulfur particles into the stratosphere [CRU 06]. Not only would these techniques not offer any solution to the problem of acidification, but they would also allow sustained CO_2 emissions with limiting consequences on the Earth's temperature. Furthermore, these techniques would modify the highly complex processes and feedback of the climate system with completely unknown secondary effects [WIL 12].

In conclusion, although CDR techniques do present interesting possibilities by directly addressing the cause of ocean acidification (in contrast to SRM techniques), the cost of their implementation and the CO_2 consumption that some need would limit their effectiveness. Williamson and Turley [WIL 12] predicted that they would allow for a maximum CO_2 reduction of around 10–50% of the actual emissions and that there is a risk associated with a few of them. Geoengineering techniques can, therefore, only be seen as a partial solution which would complement a reduction of CO_2 emissions.

5.5.3. *Adaptation measures*

Adaptation methods, often used on a local or regional level following a precise and localized environmental impact, can increase the resilience of marine ecosystems by reducing other stress factors and thus slowing down the harmful effects of ocean acidification. These methods could be a quicker and more acceptable option for decision makers than just CO_2 mitigation [BIL 13]. For example, coastal pollution due to the delivery of phosphate and nitrate from agriculture or urban waste is one of the causes of ocean acidification in certain regions. Limiting this pollution can, therefore, contribute toward tackling ocean acidification [KEL 11]. The creation of marine protected areas is a well-known method, which has already been implemented with success for addressing other problems. Nevertheless, Rau *et al.* [RAU 12] are concerned that the impact on marine-protected areas would be too limited and slow for efficiently fighting off the harmful effects of acidification and global warming on corals. In the specific case of the Whiskey Creek Fish Hatchery in the United States, adaptation methods have been implemented very quickly. The hatchery regulates the supply of seawater to its facility depending on its acidity (lower in the afternoon due to natural daily variations), and adds calcium chloride and sodium carbonate in order to increase the saturation level of $CaCO_3$ [BAR 12]. Another, more radical, adaptation method has recently been implemented by a hatchery: the relocation of the production of larvae in regions where the seawater acidity is low. Goose Point Oysters moved part of its production from Oregon to Hawaii [BRA 13]. The spat is then transported to the Northwest American coast for growth, which does not seem wise due to the CO_2 emissions caused by transportation.

Finally, the possibility of identifying species that are either resistant or tolerant to acidification is another adaptation mechanism. Some Australian oyster populations of *Saccostrea glomerata* are more resistant than others to the drop in pH and the concentration in carbonate ions [PAR 11]. Furthermore, the larvae of this species are more resistant to elevated levels of pCO_2 when their parents have already been exposed to elevated CO_2 [PAR 12]. These results suggest that the genetic selection of strains resistant to ocean acidification is possible.

As with *carbon dioxide removal techniques*, adaptation methods should not be considered as an alternative to the reduction of CO_2 emissions but as additional measures aiming to offer temporary protection to marine organism ecosystems.

5.6. Conclusion

Some effects of CO_2 uptake by the oceans, for example, changes in the carbonate chemistry, are known with almost complete certainty [GAT 11c, GAT 13]. The majority of the effects on the biology and ecology of organisms and ecosystems are less well known. There is no doubt, however, that calcification, primary production, nitrogen fixation and diversity will be modified, some positively, others negatively, but with unknown magnitude. These changes will have repercussions on biogeochemical cycles and socio-economic activities, even though it is currently impossible to determine the magnitude of these consequences. These uncertainties pose a problem, since decision makers require reliable predictions on the future of the ocean which can be used in international negotiations, for example, those aiming to reduce CO_2 emissions as part of the United Nations Framework Convention on Climate Change. Although the economic and social impacts [GON 10] of climate change are already perceptible, these negotiations are making little progress. More concrete information on the consequences of the acidification of oceans could reinforce the view that progress at the negotiating table is absolutely crucial. Uncertainties, unknowns and certain contradictory results have not enabled ocean acidification to become a prominent issue in international discussions. This is probably not the only reason, it is also possible that certain decision makers feel that there are already very difficult climate change issues to address without having to consider another problem, even if climate change and ocean acidification share the same origin. Limiting CO_2 emissions would address both processes.

There is a much more optimistic way of interpreting the state of knowledge on ocean acidification. Even though we have known for several decades that the uptake of CO_2 by the ocean leads to changes in seawater chemistry, it is only recently that biological, ecological and biogeochemical consequences have been identified. If we consider that, until very recently, few researchers focused on this issue, the progress made in less than 20 years has been remarkable. The comparison with the development of research on the impact of global change on terrestrial ecosystems is not disadvantageous. It is clear that major progress will be made in the next few years due to the launch of national and international research projects, with the considerable increase in the number of researchers involved, and the assessment performed as part of the Fifth Assessment Report of the IPCC. Whether upcoming results will lead to a more positive or more dramatic assessment of the severity of the impact of ocean acidification, there is no doubt that the information will be most valuable for human society in deciding the best decisions to make.

5.7. Acknowledgments

This chapter is a contribution to the Mediterranean Sea Acidification in a Changing Climate (MedSeA) project, which benefits from financial support from the European Commission, and the Ocean Acidification International Coordination Centre (OA-ICC) funded through the IAEA Peaceful Uses Initiative. The IAEA is grateful for the support provided to its Enviornmental Laboratories by the government of the Principality of Monaco.

5.8. Appendix

5.8.1. *Carbonate chemistry of carbonates and biogeochemical processes*

The chemistry of seawater is relatively complex and can only be briefly summarized here. The works of Zeebe and Wolf-Gadrow [ZEE 01] and Millero [MIL 06] provide a very complete analysis on the topic.

Dissolved inorganic carbon (DIC or C_T) is present in different forms in seawater. The main forms are dissolved carbonic gas (CO_2), bicarbonate (HCO_3^-) and carbonate (CO_3^{2-}). The pH is an indicator of seawater acidity (pH = $-\log_{10}[H^+]$) which can be formulated using different scales; the total scale is used in this chapter [DIC 10]. The concentration in carbonic acid (H_2CO_3) is weak, generally less than 0.3%, and is often combined with that of dissolved CO_2 ($[CO_2]$). These different forms of carbon are in equilibrium depending on their relative concentration, pH, temperature and the concentration of other compounds (Figure 5.1). The average chemistry of carbonates in surface waters in 1766, 2007 and 2100 is shown in Table 5.1.

The three main forms of calcium carbonate ($CaCO_3$) in seawater are aragonite, calcite and magnesian calcite. Aragonite is around 1.5 times more soluble than calcite, whereas the solubility of magnesian calcite is less than that of calcite when the magnesium content is weak (less than 4%), however, it is higher when the magnesium content is above 12% [DIC 10]. The equilibrium of calcium is the following:

$$CaCO_3 \leftrightharpoons Ca^{2+}(aq) + CO_3^{2-}(aq) \quad [5.1]$$

The equilibrium constant is defined by the solubility product of calcite or aragonite:

$$K^*_{sp} = [Ca^{2+}]_{sat} \times [CO_3^{2-}]_{sat} \quad [5.2]$$

The saturation of $CaCO_3$ is defined by the ratio of the ionic product measured with the ionic product expected for a solution that is in equilibrium with calcite or aragonite:

$$\Omega = \frac{[Ca^{2+}] \times [CO_3^{2-}]}{K^*_{sp}} \quad [5.3]$$

Seawater is in equilibrium with the considered mineral when $\Omega = 1$, over-saturated when $\Omega > 1$ (favors the precipitation of $CaCO_3$) and under-saturated when $\Omega < 1$ (favors the dissolution of $CaCO_3$).

CO_2 becomes a weak acid (carbonic acid) when it dissolves in seawater. It increases the concentration of HCO_3^- and C_T and reduces the pH, the concentration of CO_3^{2-} and the saturation of calcium carbonate:

$$CO_2 + CO_3^{2-} + H_2O \rightarrow 2HCO_3^- \quad [5.4]$$

Many processes, for example, enzymatic activity [MAD 88], depend on the pH. The two biological processes, which have been the most studied within the context of ocean acidification, are the precipitation and dissolution of $CaCO_3$ (equation [5.5]) as well as photosynthesis and respiration (equation [5.6]):

$$Ca^{2+} + 2HCO_3^- \leftrightharpoons 6\,CaCO_3 + CO_2 + H_2O \quad [5.5]$$

$$106CO_2 + 16NO_3^- + HPO_4^{2-} + 122H_2O + 18H^+ \leftrightharpoons 5$$
$$C_{106}H_{263}O_{110}N_{16}P + 138O_2 \quad [5.6]$$

5.9. Bibliography

[AGE 85] AGEGIAN C.R., The biogeochemical ecology of *Porolithon gardineri* (Foslie), PhD Thesis, University of Hawaii, 1985.

[ALL 08] ALLGAIER M., RIEBESELL U., VOGT M., *et al.*, "Coupling of heterotrophic bacteria to phytoplankton bloom development at different pCO_2 levels: a mesocosm study", *Biogeosciences*, vol. 5, pp. 1007–1022, 2008.

[AND 05] ANDERSSON A.J., MACKENZIE F.T., LERMAN A., "Coastal ocean and carbonate systems in the high CO_2 world of the anthropocene", *American Journal of Science*, vol. 305, pp. 875–918, 2005.

[AND 11] ANDERSSON A.J., MACKENZIE F.T., GATTUSO J.-P., "Effects of ocean acidification on benthic processes, organisms, and ecosystems", in GATTUSO J.-P., HANSSON L. (eds.), *Ocean Acidification*, Oxford University Press, Oxford, pp. 122–153, 2011.

[ANT 08] ANTHONY K.R.N., KLINE D.L., DIAZ-PULIDO G., *et al.*, "Ocean acidification causes bleaching and productivity loss in coral reef builders", *Proceedings of the National Academy of Science USA*, vol. 105, pp. 17442–17446, 2008.

[ANT 11] ANTHONY K.N., MAYNARD J.A., DIAZ-PULIDO G., *et al.*, "Ocean acidification and warming will lower coral reef resilience", *Global Change Biology*, vol. 17, pp. 1798–1808, 2011.

[ARN 12] ARNOLD T., MEALEY C., LEAHEY H., *et al.*, "Ocean acidification and the loss of phenolic substances in marine plants", *PLoS ONE,* vol. 7, p. e35107, 2012.

[AZA 83] AZAM F., FENCHEL T., FIELD J.G., *et al.*, "The ecological role of water-column microbes in the sea", *Marine Ecology Progress Series*, vol. 10, pp. 257–263, 1983.

[BAR 02] BARKER S., ELDERFIELD H., "Foraminiferal calcification response to glacial-interglacial changes in atmospheric CO_2", *Science*, vol. 297, pp. 833–836, 2002.

[BAR 03] BARKER S., HIGGINS J.A., ELDERFIELD H., "The future of the carbon cycle: review, calcification response, ballast and feedback on atmospheric CO_2", *Philosophical Transactions of the Royal Society of London Series A*, vol. 361, no. 1810, pp. 1977–1999, 2003.

[BAR 12] BARTON A., HALES B., WALDBUSSER G.G., *et al.*, "The Pacific oyster, *Crassostrea gigas*, shows negative correlation to naturally elevated carbon dioxide levels: implications for near-term ocean acidification effects", *Limnology and Oceanography*, vol. 57, pp. 698–710, 2012.

[BAT 01] BATES N.R., "Interannual variability of oceanic CO_2 and biogeochemical properties in the Western North Atlantic subtropical gyre", *Deep-Sea Research* (*Part II, Topical Studies in Oceanography*), vol. 48, pp. 1507–1528, 2001.

[BAT 07] BATES N.R., "Interannual variability of the oceanic CO_2 sink in the subtropical gyre of the North Atlantic Ocean over the last 2 decades", *Journal of Geophysical Research Oceans*, vol. 112, p. C09013, 2007.

[BAU 11] BAUMANN H., TALMAGE S.C., GOBLER C.J., "Reduced early life growth and survival in a fish in direct response to increased carbon dioxide", *Nature Climate Change*, vol. 2, pp. 38–41, 2011.

[BAY 76] BAYNE B.L., *Marine Mussels: Their Ecology and Physiology*, Cambridge University Press, Cambridge, 1976.

[BEA 11] BEAUFORT L., PROBERT I., DE GARIDEL-THORON T., *et al.*, "Sensitivity of coccolithophores to carbonate chemistry and ocean acidification", *Nature*, vol. 476, pp. 80–83, 2011.

[BED 12] BEDNARŠEK N., TARLING G.A., BAKKER D.C.E., *et al.*, "Extensive dissolution of live pteropods in the Southern Ocean", *Nature Geoscience*, vol. 5, pp. 881–885, 2012.

[BIL 12] BILLÉ R., KELLY R., BIASTOCH A., *et al.*, Que faire face à l'acidification des océans?, IDDRI-Sciences Po Policy Brief N°17/12 8, IDDRI-Sciences Po, 2012.

[BIL 13] BILLÉ R., KELLY R., BIASTOCH A., *et al.*, "Taking action against ocean acidification: a review of management and policy options", *Environmental Management*, vol. 52, pp. 761–779, 2013.

[BOR 11] BORGES A.V., "Present day carbon dioxide fluxes in the coastal ocean and possible feedbacks under global change", in DA SILVA DUARTE P.M., SANTANA-CASIANO J.M. (eds.), *Oceans and the atmospheric carbon content*, Dordrecht, Springer, pp. 47–77, 2011.

[BOU 26] BOUXIN H., "Action des acides sur le squelette des larves de l'oursin *Paracentrotus lividus*", *Influence du pH. Comptes Rendus des Séances de la Société de Biologie*, vol. 94, pp. 453–455, 1926.

[BRA 13] BRANCH T.A., DEJOSEPH B.M., RAY L.J., *et al.*, "Impacts of ocean acidification on marine seafood", *Trends in Ecology & Evolution*, vol. 28, pp. 178–186, 2013.

[BRA 12] BRANDER L.M., REHDANZ K., TOL R.S.J., *et al.*, "The economic impact of ocean acidification on coral reefs", *Climate Change Economics*, vol. 03, p. 1250002, 2012.

[BRE 78] BREWER P.G., "Direct observation of the oceanic CO_2 increase", *Geophysical Research Letters*, vol. 5, pp. 997–1000, 1978.

[BRO 66] BROECKER W.S., TAKAHASHI T., "Calcium carbonate precipitation on the Bahama Banks", *Journal of Geophysical Research*, vol. 71, pp. 1575–1602, 1966.

[BRO 01] BROECKER W.S., CLARK E., "A dramatic Atlantic dissolution event at the onset of the last glaciation", *Geochemistry Geophysics Geosystems*, vol. 2, 2001.

[CAI 11] CAI W.J., HU X., HUANG W.J., *et al.*, "Acidification of subsurface coastal waters enhanced by eutrophication", *Nature Geoscience*, vol. 4, pp. 766–770, 2011.

[CAL 03] CALDEIRA K., WICKETT M.E., "Anthropogenic carbon and ocean pH", *Nature*, vol. 425, p. 365, 2003.

[CAM 11] CAMPBELL J.E., FOURQUREAN J.W., "Novel methodology for *in situ* carbon dioxide enrichment of benthic ecosystems", *Limnology and Oceanography: Methods*, vol. 9, pp. 97–109, 2011.

[CAO 10] CAO L., CALDEIRA K., "Can ocean iron fertilization mitigate ocean acidification?", *Climatic Change*, vol. 99, pp. 303–311, 2010.

[CAR 07] CAROTENUTO Y., PUTZEYS S., SIMONELLI P., *et al.*, "Copepod feeding and reproduction in relation to phytoplankton development during the PeECE III mesocosm experiment", *Biogeosciences Discussions*, vol. 4, pp. 3913–3936, 2007.

[CES 03] CESAR H.S.J., BURKE L.M., PET-SOEDE L., *The economics of worldwide coral reef degradation*, Cesar Environmental Economics Consulting (CEEC), Arnhem, The Netherlands, 2003.

[CHE 09] CHECKLEY D.M., DICKSON A.G., TAKAHASHI M., *et al.*, "Elevated CO_2 enhances otolith growth in young fish", *Science*, vol. 324, pp. 1683–1683, 2009.

[COM 12] COMEAU S., GATTUSO J.P., JEFFREE R., *et al.*, "Effect of carbonate chemistry manipulations on calcification, respiration, and excretion of a Mediterranean pteropod", *Biogeosciences Discussions*, vol. 9, pp. 6169–6189, 2012.

[COM 09] COMEAU S., GORSKY G., JEFFREE R., *et al.*, "Impact of ocean acidification on a key Arctic pelagic mollusc (Limacina helicina)", *Biogeosciences*, vol. 6, pp. 1877–1882, 2009.

[COM 10a] COMEAU S., GORSKY G., ALLIOUANE S., *et al.*, "Larvae of the pteropod Cavolinia inflexa exposed to aragonite undersaturation are viable but shell-less", *Marine Biology*, vol. 157, pp. 2341–2345, 2010.

[COM 10b] COMEAU S., JEFFREE R., TEYSSIÉ J.L., *et al.*, "Response of the Arctic pteropod *Limacina helicina* to projected future environmental conditions", *PLoS ONE*, vol. 5, p. e11362, 2010.

[COO 09] COOLEY S.R., DONEY S.C., "Anticipating ocean acidification's economic consequences for commercial fisheries", *Environmental Research Letters*, vol. 4, pp. 1–8, 2009.

[CRU 06] CRUTZEN P.J., "Albedo enhancement by stratospheric sulfur injections: a contribution to resolve a policy dilemma?", *Climatic Change*, vol. 77, pp. 211–219, 2006.

[CZE 09] CZERNY J., RAMOS J.B.E., RIEBESELL U., "Influence of elevated CO_2 concentrations on cell division and nitrogen fixation rates in the bloom-forming cyanobacterium *Nodularia spumigena*", *Biogeosciences*, vol. 6, pp. 1865–1875, 2009.

[DEN 07] DENMAN K.L., BRASSEUR G., CHIDTHAISONG A., *et al.*, "Couplings between changes in the climate system and biogeochemistry", in SOLOMON S., QIN D., MANNING M. (eds.), *Climate Change 2007: The Physical Science Basis. Contribution of Working Group I to the Fourth Assessment Report of the Intergovernmental Panel on Climate Change*, Cambridge University Press, Cambridge pp. 499–587, 2007.

[DIA 12] DIAZ-PULIDO G., ANTHONY K.R.N., KLINE D.I., *et al.*, "Interactions between ocean acidification and warming on the mortality and dissolution of coralline algae", *Journal of Phycology*, vol. 48, pp. 32–39, 2012.

[DIC 10] DICKSON A., "The carbon dioxide system in seawater: equilibrium chemistry and measurements", in RIEBESELL U., FABRY V.J., HANSSON L. (eds.), *Guide to Best Practices for Ocean Acidification Research and Data Reporting*, Publications Office of the European Union, Luxembourg, pp. 17–40, 2010.

[DON 07] DONEY S.C., MAHOWALD N., LIMA I., *et al.*, "Impact of anthropogenic atmospheric nitrogen and sulfur deposition on ocean acidification and the inorganic carbon system", *Proceedings of the National Academy of Science USA*, vol. 104, pp. 14580–14585, 2007.

[DOR 09] DORE J.E., LUKAS R., SADLER D.W., *et al.*, "Physical and biogeochemical modulation of ocean acidification in the central North Pacific", *Proceedings of the National Academy of Science USA*, vol. 106, pp. 12235–12240, 2009.

[DUP 10] DUPONT S., ORTEGA-MARTÍNEZ O., THORNDYKE M., "Impact of near-future ocean acidification on echinoderms", *Ecotoxicology*, vol. 19, pp. 449–462, 2010.

[ENG 05] ENGEL A., ZONDERVAN I., AERTS K., *et al.*, "Testing the direct effect of CO_2 concentration on a bloom the coccolithophorid *Emiliania huxleyi* in mesocosm experiments", *Limnology and Oceanography*, vol. 50, pp. 493–507, 2005.

[ESB 12] ESBAUGH A.J., HEUER R., GROSELL M., "Impacts of ocean acidification on respiratory gas exchange and acid–base balance in a marine teleost, *Opsanus beta*", *Journal of Comparative Physiology B: Biochemical, Systemic, and Environmental Physiology*, vol. 182, pp. 921–934, 2012.

[FAO 09] FAO, The state of fisheries and aquaculture 2008, 196, Food and Agriculture Organization of the United Nations, Rome, Italy, 2009.

[FEE 08] FEELY R.A., SABINE C.L., HERNANDEZ-AYON J.M., *et al.*, "Evidence for upwelling of corrosive "acidified" water onto the continental shelf", *Science*, vol. 320, pp. 1490–1492, 2008.

[FEN 09] FENG Y., HARE C.E., LEBLANC K., *et al.*, "Effects of increased pCO_2 and temperature on the North Atlantic spring bloom. I. The phytoplankton community and biogeochemical response", *Marine Ecology Progress Series*, vol. 388, pp. 13–25, 2009.

[FIE 98] FIELD C.B., BEHRENFELD M.J., RANDERSON J.T., *et al.*, "Primary production of the biosphere: integrating terrestrial and oceanic components", *Science*, vol. 281, no. 5374, pp. 237–240, 1998.

[FOR 12] FORM A.U., RIEBESELL U., "Acclimation to ocean acidification during long-term CO_2 exposure in the cold-water coral *Lophelia pertusa*", *Global Change Biology*, vol. 18, pp. 843–853, 2012.

[FRA 94] FRANKIGNOULLE M., CANON C., GATTUSO J.-P., "Marine calcification as a source of carbon dioxide – positive feedback of increasing atmospheric CO_2", *Limnology and Oceanography*, vol. 39, pp. 458–462, 1994.

[FRI 13] FRIELER K., MEINSHAUSEN M., GOLLY A., *et al.*, "Limiting global warming to 2°C is unlikely to save most coral reefs", *Nature Climate Change*, vol. 3, pp. 165–170, 2013.

[FRO 12] FROMMEL A.Y., MANEJA R., LOWE D., *et al.*, "Severe tissue damage in Atlantic cod larvae under increasing ocean acidification", *Nature Climate Change*, vol. 2, pp. 42–46, 2012.

[FU 08] FU F.X., MULHOLLAND M.R., GARCIA N.S., *et al.*, "Interactions between changing pCO_2, N_2 fixation, and Fe limitation in the marine unicellular cyanobacterium *Crocosphaera*", *Limnology and Oceanography*, vol. 53, pp. 2472–2484, 2008.

[FU 10] FU F.X., PLACE A.R., GARCIA N.S., *et al.*, "CO_2 and phosphate availability control the toxicity of the harmful bloom dinoflagellate *Karlodinium veneficum*", *Aquatic Microbial Ecology*, vol. 59, pp. 55–65, 2010.

[GAI 19] GAIL F.W., "Hydrogen ion concentration and other factors affecting the distribution of *Fucus*", *Publication Puget Sound Biological Station*, vol. 2, pp. 287–306, 1919.

[GAO 93] GAO K.S., ARUGA Y., ASADA K., *et al.*, "Influence of enhanced CO_2 on growth and photosynthesis of the red algae *Gracilaria* sp. and *G. chilensis*", *Journal of Applied Phycology*, vol. 5, no. 6, pp. 563–571, 1993.

[GAO 09] GAO K.S., RUAN Z., VILLAFAÑE V.E., *et al.*, "Ocean acidification exacerbates the effect of UV radiation on the calcifying phytoplankter *Emiliania huxleyi*", *Limnology and Oceanography*, vol. 54, pp. 1855–1862, 2009.

[GAR 06] GARCIA H.E., LOCARNINI R.A., BOYER T.P., *et al.*, "World Ocean Atlas 2005, Volume 4: Nutrients (phosphate, nitrate, silicate)", in LEVITUS S., (ed.), *NOAA Atlas NESDIS*, U.S. Government Printing Office, Washington, D.C., vol. 64, pp. 1–396, 2006.

[GAT 98] GATTUSO J.P., FRANKIGNOULLE M., BOURGE I., *et al.*, "Effect of calcium carbonate saturation of seawater on coral calcification", *Global and Planetary Change*, vol. 18, pp. 37–46, 1998.

[GAT 09] GATTUSO J.P., LAVIGNE H., "Technical note: approaches and software tools to investigate the impact of ocean acidification", *Biogeosciences*, vol. 6, pp. 2121–2133, 2009.

[GAT 11a] GATTUSO J.P., HANSSON L., "Ocean acidification: background and history", in GATTUSO J.-P., HANSSON L. (eds.), *Ocean acidification*, pp. 1–20, Oxford University Press, Oxford, 2011.

[GAT 11b] GATTUSO J.P., RIEBESELL U., "Reconciling apparently contradictory observations", in FIELD C.B., BARROS V., STOCKER T.F., *et al.*, (eds.), *Workshop report of the Intergovernmental Panel on Climate Change workshop on impacts of ocean acidification on marine biology and ecosystems*, pp. 10–16, IPCC Working Group II Technical Support Unit, Carnegie Institution, Stanford, TN, 2011.

[GAT 11c] GATTUSO J.P., BIJMA J., GEHLEN M., *et al.*, "Ocean acidification: knowns, unknowns and perspectives", in GATTUSO J.-P., HANSSON L., (eds.), *Ocean Acidification*, pp. 291–311, Oxford University Press, Oxford, 2011.

[GAT 13] GATTUSO J.P., MACH K.J., MORGAN G.M., "Ocean acidification and its impacts: an expert survey", *Climatic Change*, vol. 117, pp. 725–738, 2013.

[GAZ 07] GAZEAU F., QUIBLIER C., JANSEN J.M., *et al.*, "Impact of elevated CO_2 on shellfish calcification", *Geophysical Research Letters*, vol. 34, p. L07603, 2007.

[GAZ 10] GAZEAU F., GATTUSO J.P., DAWBER C., *et al.*, "Effect of ocean acidification on the early life stages of the blue mussel (*Mytilus edulis*)", *Biogeosciences*, vol. 7, pp. 2051–2060, 2010.

[GEH 07] GEHLEN M., GANGSTØ R., SCHNEIDER B., *et al.*, "The fate of pelagic $CaCO_3$ production in a high CO_2 ocean: A model study", *Biogeosciences*, vol. 4, pp. 505–519, 2007.

[GEH 11] GEHLEN M., GRUBER N., GANGSTØ R., *et al.*, "Biogeochemical consequences of ocean acidification and feedback to the Earth system", in GATTUSO J.-P., HANSSON L. (eds.), *Ocean Acidification*, Oxford University Press, Oxford, pp. 230–248, 2011.

[GON 10] GONZÁLEZ-DAVILÁ M., SANTANA-CASIANO J.M., RUEDA M.J., *et al.*, "The water column distribution of carbonate system variables at the ESTOC site from 1995 to 2004", *Biogeosciences*, vol. 7, pp. 3067–3081, 2010.

[GRO 06] GROSSART H.-P., ALLGAIER M., PASSOW U., *et al.*, "Testing the effect of CO_2 concentration on the dynamics of marine heterotrophic bacterioplankton", *Limnology and Oceanography*, vol. 51, pp. 1–11, 2006.

[HAL 11] HALE R., CALOSI P., MCNEILL L., *et al.*, "Predicted levels of future ocean acidification and temperature rise could alter community structure and biodiversity in marine benthic communities", *Oikos*, vol. 120, pp. 661–674, 2011.

[HAL 08] HALL-SPENCER J.M., RODOLFO-METALPA R., MARTIN S., *et al.*, "Volcanic carbon dioxide vents show ecosystem effects of ocean acidification", *Nature*, vol. 454, pp. 96–99, 2008.

[HAL 93] HALLEGRAEFF G.M., "A review of harmful algal blooms and their apparent global increase", *Phycologia*, vol. 32, pp. 79–99, 1993.

[HAM 12] HAMMOND L.M., HOFMANN G.E., "Early developmental gene regulation in *Strongylocentrotus purpuratus* embryos in response to elevated CO_2 seawater conditions", *Journal of Experimental Biology*, vol. 215, pp. 2445–2454, 2012.

[HAR 07] HARE C.E., LEBLANC K., DITULLIO G.R., *et al.*, "Consequences of increased temperature and CO_2 for phytoplankton community structure in the Bering Sea", *Marine Ecology Progress Series*, vol. 352, pp. 9–16, 2007.

[HAR 12] HARROULD-KOLIEB E.R., HERR D., "Ocean acidification and climate change: synergies and challenges of addressing both under the UNFCCC", *Climate Policy*, vol. 12, pp. 378–389, 2012.

[HAV 08] HAVENHAND J.N., BUTTLER F.R., THORNDYKE M.C., *et al.*, "Near future levels of ocean acidification reduce fertilization success in a sea urchin", *Current Biology*, vol. 18, pp. R651–R652, 2008.

[HEI 04] HEINZE C., "Simulating oceanic $CaCO_3$ export production in the greenhouse", *Geophysical Research Letters*, vol. 31, p. L16308, 2004.

[HEN 10] HENDRIKS I.E., DUARTE C.M., ALVAREZ M., "Vulnerability of marine biodiversity to ocean acidification: a meta-analysis", *Estuarine, Coastal and Shelf Science*, vol. 86, pp. 157–164, 2010.

[HER 11] HERNROTH B., BADEN S., THORNDYKE M., *et al.*, "Immune suppression of the echinoderm *Asterias rubens* (L.) following long-term ocean acidification", *Aquatic Toxicology*, vol. 103, pp. 222–224, 2011.

[HIL 13] HILMI N., ALLEMAND D., DUPONT S., *et al.*, "Towards improved socio-economic assessments of ocean acidification's impacts", *Marine Biology*, vol. 160, pp. 1773–1787, 2013.

[HOE 11] HOEGH-GULDBERG O., "Coral reef ecosystems and anthropogenic climate change", *Regional Environmental Change*, vol. 11, pp. 215–227, 2011.

[HON 12] HÖNISCH B., RIDGWELL A., SCHMIDT D.N., *et al.*, "The geological record of ocean acidification", *Science*, vol. 335, pp. 1058–1063, 2012.

[HUT 07] HUTCHINS D.A., FU F.X., ZHANG Y., *et al.*, "CO_2 control of *Trichodesmium* N_2 fixation, photosynthesis, growth rates, and elemental ratios: Implications for past, present, and future ocean biogeochemistry", *Limnology and Oceanography*, vol. 52, pp. 1293–1304, 2007.

[IAE 11] Economics of ocean acidification – bridging the gap between ocean acidification impacts and economic valuation. Conclusions and recommendations for decision makers, IAEA and Centre Scientifique de Monaco, Monaco, 2011.

[IGL 08] IGLESIAS-RODRIGUEZ M.D., HALLORAN P.R., RICKABY R.E.M., *et al.*, "Phytoplankton calcification in a high-CO_2 world", *Science*, vol. 320, pp. 336–340, 2008.

[IPP 07] IPCC, Climate change 2007: the physical science basis, Contribution of Working Group I to the Fourth Assessment Report of the Intergovernmental Panel on Climate Change, Cambridge, New York, Cambridge University Press, p. 996, 2007.

[ISH 04] ISHIMATSU A., KIKKAWA T., HAYASHI M., *et al.*, "Effects of CO_2 on marine fish: larvae and adults", *Journal of Oceanography*, vol. 60, pp. 731–741, 2004.

[JOO 11] JOOS F., FRÖLICHER T.L., STEINACHER M., *et al.*, "Impact of climate change mitigation on ocean acidification projections", in GATTUSO J.-P., HANSSON L. (eds.), *Ocean Acidification*, Oxford University Press, Oxford, pp. 272–290, 2011.

[KEE 08] KEELING R.F., PIPER S.C., BOLLENBACHER A.F., *et al.*, "Atmospheric CO_2 records from sites in the SIO air sampling network", *Trends: A Compendium of Data on Global Change*, Carbon Dioxide Information Analysis Center, Oak Ridge National Laboratory, U.S. Department of Energy, Oak Ridge, TN, 2008.

[KEE 10] KEELING R.F., KÖRTZINGER A., GRUBER N., "Ocean deoxygenation in a warming world", *Annual Review of Marine Science*, vol. 2, pp. 199–229, 2010.

[KEL 11] KELLY R.P., FOLEY M.M., FISHER W.S., *et al.*, "Mitigating local causes of ocean acidification with existing laws", *Science*, vol. 332, pp. 1036–1037, 2011.

[KIM 12] KIM R.E., "Is a new multilateral environmental agreement on ocean acidification necessary?", *Review of European Community & International Environmental Law*, vol. 21, pp. 243–258, 2012.

[KIR 09] KIRKWOOD W., SANO L., "Developing new instrumentation for in situ experimentation related to ocean acidification – scaling up pH effects from the lab to the field", *Current, the Journal of Marine Education*, vol. 25, pp. 13–14, 2009.

[KLE 99] KLEYPAS J.A., BUDDEMEIER R.W., ARCHER D., *et al.*, "Geochemical consequences of increased atmospheric CO_2 on coral reefs", *Science*, vol. 284, pp. 118–120, 1999.

[KLI 12] KLINE D.I., TENEVA L., SCHNEIDER K., *et al.*, "A short-term in situ CO_2 enrichment experiment on Heron Island (GBR)", *Scientific Reports*, vol. 2, p. 413, 2012.

[KRO 10] KROEKER K., KORDAS R.L., CRIM R.N., *et al.*, "Meta-analysis reveals negative yet variable effects of ocean acidification on marine organisms", *Ecology Letters*, vol. 13, pp. 1419–1434, 2010.

[KRO 13] KROEKER K., KORDAS R.C., CRIM R., *et al.*, "Impacts of ocean acidification on marine organisms: quantifying sensitivities and interaction with warming", *Global Change Biology*, vol. 19, p. 6, 2013.

[KUF 07] KUFFNER I.B., ANDERSSON A.J., JOKIEL P.L., *et al.*, "Decreased abundance of crustose coralline algae due to ocean acidification", *Nature Geoscience*, vol. 1, pp. 114–117, 2007.

[KUR 04] KURIHARA H., SHIMODE S., SHIRAYAMA Y., "Effects of raised CO_2 concentration on the egg production rate and early development of two marine copepods (*Acartia steuri* and *Acartia erythraea*)", *Marine Pollution Bulletin*, vol. 49, nos. 9–10, pp. 721–727, 2004.

[LAF 11] LAFFOLEY D., BAXTER J.M. (eds.), Ocean acidification: acting on evidence- Messages for Rio+20, European Project on Ocean Acidification (EPOCA), UK Ocean Acidification Research Programme, (UKOA), Biological Impacts of Ocean Acidification (BIOACID) and Mediterranean Sea Acidification in a Changing Climate (MedSeA), 2011.

[LAN 00] LANGDON C., TAKAHASHI T., MARUBINI F., *et al.*, "Effect of calcium carbonate saturation state on the rate of calcification of an experimental coral reef", *Global Biogeochemical Cycles*, vol. 14, pp. 639–654, 2000.

[LAN 06] LANGER G., GEISEN M., BAUMANN K.H., *et al.*, "Species-specific responses of calcifying algae to changing seawater carbonate chemistry", *Geochemistry, Geophysics, Geosystems*, vol. 7, p. Q09006, 2006.

[LAV 10] LAVIGNE H., GATTUSO J.-P., Seacarb: seawater carbonate chemistry with R. R package version 2.3.3, 2010.

[LEQ 09] LE QUÉRÉ C., RAUPACH M.R., CANADELL J.G., *et al.*, "Trends in the sources and sinks of carbon dioxide", *Nature Geoscience*, vol. 2, pp. 831–836, 2009.

[LEC 00] LECLERCQ N., GATTUSO J.-P., JAUBERT J., "CO_2 partial pressure controls the calcification rate of a coral community", *Global Change Biology*, vol. 6, pp. 329–334, 2000.

[LIU 10] LIU J., WEINBAUER M.G., MAIER C., *et al.*, "Effect of ocean acidification on microbial diversity, and on microbe-driven biogeochemistry and ecosystem functioning", *Aquatic Microbial Ecology*, vol. 61, pp. 291–305, 2010.

[LOM 10] LOMBARD F., DA ROCHA R.E., BIJMA J., *et al.*, "Effect of carbonate ion concentration and irradiance on calcification in planktonic foraminifera", *Biogeosciences*, vol. 7, pp. 247–255, 2010.

[LUE 00] LUEKER T.J., DICKSON A.G., KEELING C.D., "Ocean pCO_2 calculated from dissolved inorganic carbon, alkalinity, and equations for K_1 and K_2: validation based on laboratory measurements of CO_2 in gas and seawater at equilibrium", *Marine Chemistry*, vol. 70, pp. 105–119, 2000.

[MAD 88] MADSHUS I.H., "Regulation of intracellular pH in eukaryotic cells", *Biochemical Journal*, vol. 250, pp. 1–8, 1988.

[MAI 12] MAIER C., WATREMEZ P., TAVIANI M., *et al.*, "Calcification rates and the effect of ocean acidification on Mediterranean cold-water corals", *Proceedings of the Royal Society of London, Series B: Biological Sciences*, vol. 279, pp. 1716–1723, 2012.

[MAI 13] MAIER C., SCHUBERT A., BERZUNZA SÀNCHEZ M.M., *et al.*, "End of the century pCO_2 levels do not impact net calcification in Mediterranean cold-water corals", *PLoS ONE*, vol. 8, p. e62655, 2013.

[MAR 08] MARSHALL D.J., SANTOS J.H., LEUNG K.M.Y., *et al.*, "Correlations between gastropod shell dissolution and water chemical properties in a tropical estuary", *Marine Environmental Research*, vol. 66, pp. 422–429, 2008.

[MAR 08] MARTIN S., RODOLFO-METALPA R., RANSOME E., *et al.*, "Effects of naturally acidified seawater on seagrass calcareous epibionts", *Biology Letters*, vol. 4, pp. 689–692, 2008.

[MAR 09] MARTIN S., GATTUSO J.-P., "Response of Mediterranean coralline algae to ocean acidification and elevated temperature", *Global Change Biology*, vol. 15, pp. 2089–2100, 2009.

[MAR 11] MARTIN S., RICHIER S., PEDROTTI M.L., *et al.*, "Early development and molecular plasticity in the Mediterranean sea urchin *Paracentrotus lividus* exposed to CO_2-driven acidification", *The Journal of Experimental Biology*, vol. 214, pp. 1357–1368, 2011.

[MCC 18] MCCLENDON J.F., *On changes in the sea and their relation to organisms*, Carnegie Institution of Washington Publication, vol. 252, pp. 213–258, 1918.

[MIL 07] MILES H., WIDDICOMBE S., SPICER J.I., *et al.*, "Effects of anthropogenic seawater acidification on acid-base balance in the sea urchin *Psammechinus miliaris*", *Marine Pollution Bulletin*, vol. 54, pp. 89–96, 2007.

[MIL 06] MILLERO F.J., *Chemical oceanography*, CRC/Taylor and Francis, Boca Raton, 2006.

[MOO 06] MOORE B., ROAF H.E., WHITLEY E., "On the effects of alkalies and acids, and of alkaline and acid salts, upon growth and cell division in the fertilized eggs of *Echinus esculentus*. A study in relationship to the causation of malignant disease", *Proceedings of the Royal Society of London, Series B: Biological Sciences*, vol. 77, pp. 102–136, 1906.

[MOO 11] MOORE C.C., Welfare impacts of ocean acidification: an integrated assessment model of the US mollusk fishery, National Center for Environmental Economics Working Paper, 11-06, 2011.

[MOR 09] MORITA M., SUWA R., IGUCHI A., *et al.*, "Ocean acidification reduces sperm flagellar motility in broadcast spawning reef invertebrates", *Zygote*, vol. 18, pp. 103–107, 2009.

[MOY 09] MOY A.D., HOWARD W.R., BRAY S.G., *et al.*, "Reduced calcification in modern Southern Ocean planktonic foraminifera", *Nature Geoscience*, 2009.

[MUL 10] MÜLLER M.N., SCHULZ K.G., RIEBESELL U., "Effects of long-term high CO_2 exposure on two species of coccolithophores", *Biogeosciences*, vol. 7, pp. 1109–1116, 2010.

[MUN 09] MUNDAY P.L., DIXSON D.L., DONELSON J.M., *et al.*, "Ocean acidification impairs olfactory discrimination and homing ability of a marine fish", *Proceedings of the National Academy of Science USA*, vol. 106, pp. 1848–1852, 2009.

[MUN 10] MUNDAY P.L., DIXSON D.L., MCCORMICK M.I., *et al.*, "Replenishment of fish populations is threatened by ocean acidification", *Proceedings of the National Academy of Science USA*, vol. 107, pp.12930–12934, 2010.

[NAK 00] NAKIĆENOVIĆ N., ALCAMO J., DAVIS G., *et al.*, *Special report on emissions scenarios: a special report of Working Group III of the Intergovernmental Panel on Climate Change*, Cambridge University Press, Cambridge, 2000.

[NAR 12] NARITA D., REHDANZ K., TOL R.S.J., "Economic costs of ocean acidification: a look into the impacts on shellfish production", *Climatic Change*, vol. 113, pp. 1049–1063, 2012.

[NIE 13] NIEHOFF B., SCHMITHÜSEN T., KNÜPPEL N., *et al.*, "Mesozooplankton community development at elevated CO_2 concentrations: results from a mesocosm experiment in an Arctic fjord", *Biogeosciences*, vol. 10, pp. 1391–1406, 2013.

[NIL 12] NILSSON G.E., DIXSON D.L., DOMENICI P., *et al.*, "Near-future carbon dioxide levels alter fish behaviour by interfering with neurotransmitter function", *Nature Climate Change*, vol. 2, pp. 201–204, 2012.

[OCE 09] OCEAN ACIDIFICATION REFERENCE USER GROUP, "Ocean acidification: the facts. a special introductory guide for policy advisers and decision makers", in LAFFOLEY D.D.A., BAXTER J.M. (eds.), *European Project on Ocean Acidification (EPOCA)*, p. 12, 2009.

[OCE 10] OCEAN ACIDIFICATION REFERENCE USER GROUP, "Ocean acidification: questions answered", in LAFFOLEY D.D.A., BAXTER J.M. (eds.), *European Project on Ocean Acidification (EPOCA)*, p. 24, 2010.

[OLA 09] OLAFSSON J., OLAFSDOTTIR S.R., BENOIT-CATTIN A., *et al.*, "Rate of Iceland Sea acidification from time series measurements", *Biogeosciences*, vol. 6, pp. 2661–2668, 2009.

[ORR 05] ORR J.C., FABRY V.J., AUMONT O., *et al.*, "Anthropogenic ocean acidification over the twenty-first century and its impact on calcifying organisms", *Nature*, vol. 437, pp. 681–686, 2005.

[ORR 11] ORR J.C., "Recent and future changes in ocean carbonate chemistry", in GATTUSO J.-P., HANSSON L. (eds.), *Ocean Acidification*, Oxford University Press, Oxford, pp. 41–66, 2011.

[OSC 08] OSCHLIES A., SCHULZ K.G., RIEBESELL U., *et al.*, "Simulated 21st century's increase in oceanic suboxia by CO_2-enhanced biotic carbon export", *Global Biogeochemical Cycles Global Biogeochem Cycles 22*, vol. 22, pp. GB4008–GB4008, 2008.

[OSC 10] OSCHLIES A., KOEVE W., RICKELS W., *et al.*, "Side effects and accounting aspects of hypothetical large-scale Southern Ocean iron fertilization", *Biogeosciences*, vol. 7, pp. 4017–4035, 2010.

[PAL 07] PALACIOS S.L., ZIMMERMAN R.C., "Response of eelgrass *Zostera marina* to CO_2 enrichment: possible impacts of climate change and potential for remediation of coastal habitats", *Marine Ecology Progress Series*, vol. 344, pp. 1–13, 2007.

[PAR 11] PARKER L.M., ROSS P.M., O'CONNOR W.A., "Populations of the Sydney rock oyster, *Saccostrea glomerata*, vary in response to ocean acidification", *Biomedical and Life Sciences*, vol. 158, pp. 689–697, 2011.

[PAR 12] PARKER L.M., ROSS P.M., O'CONNOR W.A., *et al.*, "Adult exposure influences offspring response to ocean acidification in oysters", *Global Change Biology*, vol. 18, pp. 82–92, 2012.

[PET 12] PETERS G.P., MARLAND G., LE QUÉRÉ C., *et al.*, "Rapid growth in CO_2 emissions after the 2008-2009 global financial crisis", *Nature Climate Change*, vol. 2, pp. 2–4, 2012.

[PIO 10] PIONTEK J., LUNAU M., HÄNDEL N., *et al.*, "Acidification increases microbial polysaccharide degradation in the ocean", *Biogeosciences*, vol. 7, pp. 1615–1624, 2010.

[PLA 01] PLATTNER G.K., JOOS F., STOCKER T.F., *et al.*, "Feedback mechanisms and sensitivities of ocean carbon uptake under global warming", *Tellus B*, vol. 53, 564–592, 2001.

[POW 20] POWERS E.B., "The variation of the condition of sea-water, especially the hydrogen ion concentration, and its relation to marine organisms", *Publication Washington State University Puget Sound Biological Station*, vol. 2, pp. 369–385, 1920.

[POW 22] POWERS E.B., "The physiology of the respiration of fishes in relation to the hydrogen ion concentration of the medium", *The Journal of General Physiology*, vol. 4, pp. 305–317, 1922.

[PRI 09] Prince Albert II of Monaco Foundation, Monaco, 2009.

[PRY 29] PRYTHERCH H.F., "Investigation of the physical conditions controlling spawning of oysters and the occurrence, distribution, and setting of oyster larvae in Milford Harbor, Connecticut", *Bulletin U.S. Bureau of Fisheries*, vol. 44, pp. 429–503, 1929.

[RED 34] REDFIELD A.C., "On the proportions of organic derivatives in seawater and their relation to the composition of the plankton", in DANIEL R.J. (ed.), *James Johnstone Memorial Liverpool*, Liverpool University Press, pp. 176–192, 1934.

[RAU 12] RAU G.H., MCLEOD E.L., HOEGH-GULDBERG O., "The need for new ocean conservation strategies in a high-carbon dioxide world", *Nature Climate Change*, vol. 2, pp. 720–724, 2012.

[REI 11] REINFELDER J.R., "Carbon concentrating mechanisms in eukaryotic marine phytoplankton", *Annual Review of Marine Science*, vol. 3, pp. 291–315, 2011.

[REV 57] REVELLE R., SUESS H.E., "Carbon dioxide exchange between atmosphere and ocean and the question of an increase of atmospheric CO_2 during the past decades", *Tellus*, vol. 9, pp. 18–27, 1957.

[REY 03] REYNAUD S., LECLERCQ N., ROMAINE-LIOUD S., *et al.*, "Interacting effects of CO_2 partial pressure and temperature on photosynthesis and calcification in a scleractinian coral", *Global Change Biology*, vol. 9, pp. 1660–1668, 2003.

[RIE 00] RIEBESELL U., ZONDERVAN I., ROST B., *et al.*, "Reduced calcification of marine plankton in response to increased atmospheric CO_2", *Nature*, vol. 407, pp. 364–367, 2000.

[RIE 07] RIEBESELL U., SCHULZ K.G., BELLERBY R.G.J., *et al.*, "Enhanced biological carbon consumption in a high CO_2 ocean", *Nature*, vol. 450, pp. 545–549, 2007.

[RIE 08] RIEBESELL U., BELLERBY R.G.J., GROSSART H.P., *et al.*, "Mesocosm CO_2 perturbation studies: from organism to community level", *Biogeosciences*, vol. 5, pp. 1157–1164, 2008.

[RIE 09] RIES J.B., COHEN A.L., MCCORKLE D.C., "Marine calcifiers exhibit mixed responses to CO_2-induced ocean acidification", *Geology*, vol. 37, pp. 1131–1134, 2009.

[RIE 11] RIEBESELL U., TORTELL P.D., "Effects of ocean acidification on pelagic organisms and ecosystems", in GATTUSO J.-P., HANSSON L. (eds.), *Ocean Acidification*, pp. 99–121, Oxford University Press, Oxford, 2011.

[RIE 13] RIEBESELL U., CZERNY J., VON BRÖCKEL K., *et al.*, "Technical Note: A mobile sea-going mesocosm system – new opportunities for ocean change research", *Biogeosciences*, vol. 10, pp. 1835–1847, 2013.

[RIN 02] RINGWOOD A.H., KEPPLER C.J., "Water quality variation and clam growth: is pH really a non-issue in estuaries?", *Estuaries*, vol. 25, pp. 901–907, 2002.

[ROD 10] RODOLFO-METALPA R., MARTIN S., FERRIER-PAGÈS C., *et al.*, "Response of the temperate coral *Cladocora caespitosa* to mid- and long-term exposure to pCO_2 and temperature levels projected for the 2100 AD", *Biogeosciences*, vol. 7, pp. 289–300, 2010.

[ROD 11] RODOLFO-METALPA R., HOULBRÈQUE F., TAMBUTTÉ E., *et al.*, "Coral and mollusc resistance to ocean acidification adversely affected by warming", *Nature Climate Change*, vol. 1, pp. 308–312, 2011.

[ROS 06] ROST B., RIEBESELL U., SUELTEMEYER D., "Carbon acquisition of marine phytoplankton: Effect of photoperiod length", *Limnology and Oceanography*, vol. 51, pp. 12–20, 2006.

[SAB 04] SABINE C.L., FEELY R.A., GRUBER N., *et al.*, "The oceanic sink for anthropogenic CO_2", *Science*, vol. 305, pp. 367–371, 2004.

[SAN 07] SANTANA-CASIANO J.M., GONZÁLEZ-DÁVILA M., RUEDA M.J., *et al.*, "The interannual variability of oceanic CO_2 parameters in the northeast Atlantic subtropical gyre at the ESTOC site", *Global Biogeochemical Cycles*, vol. 21, p. GB1015, 2007.

[SAR 98] SARMIENTO J.L., HUGHES T.M.C., STOUFFER R.J., *et al.*, "Simulated response of the ocean carbon cycle to anthropogenic climate warming", *Nature*, vol. 393, pp. 245–249, 1998.

[SCH 02] SCHIEBEL R., "Planktic foraminiferal sedimentation and the marine calcite budget", *Global Biogeochemical Cycles*, vol. 16, p. 1062, 2002.

[SHI 09] SHI D., XU Y., MOREL F.M.M., "Effects of the pH/pCO_2 control method on medium chemistry and phytoplankton growth", *Biogeosciences*, vol. 6, pp. 1199–1207, 2009.

[SHI 05] SHIRAYAMA Y., THORNTON H., "Effect of increased atmospheric CO_2 on shallow water marine benthos", *Journal of Geophysical Research-Oceans*, vol. 110, p. C09S08, 2005.

[SIM 11] SIMPSON S.D., MUNDAY P.L., WITTENRICH M.L., *et al.*, "Ocean acidification erodes crucial auditory behaviour in a marine fish", *Biology Letters*, vol. 7, pp. 917–920, 2011.

[SMI 74] SMITH S.V., PESRET F., "Processes of carbon dioxide flux in the Fanning Island lagoon", *Pacific Science*, vol. 28, pp. 225–245, 1974.

[SOR 09] SØRENSEN S.P.L., "Etudes enzymatiques. II. Sur la mesure et l'importance de la concentration des ions hydrogène dans les réactions enzymatiques", *Comptes-rendus des travaux du laboratoire Carlsberg*, vol. 8, pp. 1–168, 1909.

[SPI 93] SPIVACK A.J., YOU C.F., SMITH H.J., "Foraminiferal boron isotope ratios as a proxy for surface ocean pH over the past 21 Myr", *Nature*, vol. 363, pp. 149–151, 1993.

[STE 09] STEINACHER M., JOOS F., FRÖLICHER T.L., *et al.*, "Imminent ocean acidification projected with the NCAR global coupled carbon cycle-climate model", *Biogeosciences*, vol. 6, pp. 515–533, 2009.

[SUF 08] SUFFRIAN K., SIMONELLI P., NEJSTGAARD J.C., *et al.*, "Microzooplankton grazing and phytoplankton growth in marine mesocosms with increased CO_2 levels", *Biogeosciences*, vol. 5, pp. 1145–1156, 2008.

[SUM 11] SUMAILA U.R., CHEUNG W.W.L., LAM V.W.Y., *et al.*, "Climate change impacts on the biophysics and economics of world fisheries", *Nature Climate Change*, vol. 1, pp. 449–456, 2011.

[SWI 66] SWIFT E., TAYLOR W.R., "The effect of pH on the division rate of the coccolithophorid *Cricosphaera elongata*", *Journal of Phycology*, vol. 2, pp. 121–125, 1966.

[TAL 11] TALMAGE S.C., GOBLER C.J., "Effects of elevated temperature and carbon dioxide on the growth and survival of larvae and juveniles of three species of Northwest Atlantic bivalves", *PLoS ONE*, vol. 6, p. e26941, 2011.

[THE 09] THE ROYAL SOCIETY, *Geoengineering the Climate: Science, Governance and Uncertainty*, vol. 82, The Royal Society, London, 2009.

[TOR 08] TORTELL P.D., PAYNE C.D., LI Y.Y., *et al.*, "CO_2 sensitivity of Southern Ocean phytoplankton", *Geophysical Research Letters*, vol. 35, p. ARTN L04605, 2008.

[TUN 09] TUNNICLIFFE V., DAVIES K.T.A., BUTTERFIELD D.A., *et al.*, "Survival of mussels in extremely acidic waters on a submarine volcano", *Nature Geoscience*, vol. 2, pp. 344–348, 2009.

[TUR 11] TURLEY C., BOOT K., "The ocean acidification challenges facing science and society", in GATTUSO J.-P., HANSSON L. (eds.), *Ocean Acidification*, Oxford University Press, Oxford, pp. 249–271, 2011.

[TUR 12] TURLEY C., GATTUSO J.-P., "Future biological and ecosystem impacts of ocean acidification and their socioeconomic-policy implications", *Current Opinion in Environmental Sustainability*, vol. 4, pp. 278–286, 2012.

[VAN 12] VAN COLEN C., DEBUSSCHERE E., BRAECKMAN U., *et al.*, "The early life history of the clam *Macoma balthica* in a high CO_2 world", *PLoS ONE*, vol. 7, p. e44655, 2012.

[VAN 11] VAN VUUREN D.P., EDMONDS J., KAINUMA M., *et al.*, "The representative concentration pathways: an overview", *Climatic Change*, vol. 109, pp. 5–31, 2011.

[WAN 12] WANNICKE N., ENDRES S., ENGEL A., *et al.*, "Response of *Nodularia spumigena* to pCO_2 – Part I: Growth, production and nitrogen cycling", *Biogeosciences*, vol. 9, pp. 2973–2988, 2012.

[WAT 12] WATSON S.A., PECK L.S., TYLER P.A., *et al.*, "Marine invertebrate skeleton size varies with latitude, temperature, and carbonate saturation: implications for global change and ocean acidification", *Global Change Biology*, vol. 18, pp. 3026–3038, 2012.

[WEI 11] WEINBAUER M.G., MARI X., GATTUSO J.-P., "Effect of ocean acidification on the diversity and activity of heterotrophic marine microorganisms", in GATTUSO J.-P., HANSSON L. (eds.), *Ocean Acidification*, Oxford University Press, Oxford, pp. 83–98, 2011.

[WID 11] WIDDICOMBE S., SPICER J.I., KITIDIS V., "Effects of ocean acidification on sediment fauna", in GATTUSO J.-P., HANSSON L. (eds.), *Ocean Acidification*, Oxford University Press, Oxford, pp. 176–191, 2011.

[WIL 12] WILLIAMSON P., TURLEY C., "Ocean acidification in a geoengineering context", *Philosophical Transactions of the Royal Society A*, vol. A370, pp. 4317–4342, 2012.

[WOO 08] WOOD H.L., SPICER J.I., WIDDICOMBE S., "Ocean acidification may increase calcification rates, but at a cost", *Proceedings of the Royal Society of London, Series B: Biological Sciences*, vol. 275, pp. 1767–1773, 2008.

[WU 10] WU Y., GAO K., RIEBESELL U., "CO_2-induced seawater acidification affects physiological performance of the marine diatom *Phaeodactylum tricornutum*", *Biogeosciences*, vol. 7, pp. 2915–2923, 2010.

[YU 07] YU J.M., ELDERFIELD H., HONISCH B., "B/Ca in planktonic foraminifera as a proxy for surface seawater pH", *Paleoceanography*, vol. 22, p. PA2202, 2007.

[ZEE 01] ZEEBE R.E., WOLF-GLADROW D.A., *CO_2 in Seawater: Equilibrium, Kinetics, Isotopes*, Elsevier, Amsterdam, 2001.

[ZEE 03] ZEEBE R.E., WESTBROEK P., "A simple model for the $CaCO_3$ saturation state of the ocean: the "Strangelove", the "Neritan", and the "Cretan" Ocean", *Geochemistry Geophysics Geosystems*, vol. 4, p. 1104, 2003.

[ZEE 11] ZEEBE R.E., RIDGWELL A., "Past changes of ocean carbonate chemistry", in GATTUSO J.-P., HANSSON L. (eds.), *Ocean Acidification*, Oxford University Press, Oxford, pp. 21–40, 2011.

List of Authors

Philippe BERTRAND
EPOC
University of Bordeaux 1
Talence
France

Pascale DELECLUSE
Météo-France
Saint-Mandé
and
CNRS-INSU
Paris
France

Laurence EYMARD
LOCEAN/IPSL
UPMC
Paris
France

Jean-Pierre GATTUSO
LOV
CNRS-INSU
Villefranche-sur-Mer
France

Frédéric GAZEAU
LOV
CNRS-INSU
Villefranche-sur-Mer

Lina HANSSON
Environment Laboratories
IAEA
Monaco

Louis LEGENDRE
LOV
UPMC
Villefranche-sur-Mer
France

André MARIOTTI
University Pierre and Marie Curie
Paris
France

André MONACO
CEFREM
UPVD
Perpignan
France

Jean-Charles POMEROL
University Pierre and Marie Curie
INSIS/CNRS
Paris
France

Gilles REVERDIN
LOCEAN/IPSL
UPMC
Paris
France

Patrick PROUZET
Ifremer
DS
Issy les Molineaux
France

Index

A, B

C

D, E

F, G, H

I, L, M

Other titles from

iSTE

in

Oceanography and Marine Biology

2014

Seas and Oceans Set coordinated by MONACO André, PROUZET Patrick
The Land-Sea Interactions

MUTTIN Frédéric
Marine Coastal and Water Pollutions: Oil Spill Studies

2013

LÉCUYER Christophe
Water on Earth: Physicochemical and Biological Properties

Printed and bound by CPI Group (UK) Ltd, Croydon, CR0 4YY

13/09/2026

14961077-0003